Green Energy and Technology

Manfred Stiebler

Wind Energy Systems for Electric Power Generation

With 146 Figures and 15 Tables

Author

Prof. Dr. Manfred Stiebler
Technical University of Berlin
Faculty of Electrical Engineering and
Computer Science
Inst. of Energy and Automation Technology
Einsteinufer 11, D-10587 Berlin
Germany
manfred.stiebler@iee.tu_berlin.de

ISBN: 978-3-540-68762-7   e-ISBN: 978-3-540-68765-8

Springer Series in Green Energy and Technology ISSN 1865-3529

Library of Congress Control Number: 2008929626

© 2008 Springer-Verlag Berlin Heidelberg

This work is subject to copyright. All rights are reserved, whether the whole or part of the material is concerned, specifically the rights of translation, reprinting, reuse of illustrations, recitation, broadcasting, reproduction on microfilm or in any other way, and storage in data banks. Duplication of this publication or parts thereof is permitted only under the provisions of the German Copyright Law of September 9, 1965, in its current version, and permission for use must always be obtained from Springer. Violations are liable to prosecution under the German Copyright Law.

The use of general descriptive names, registered names, trademarks, etc. in this publication does not imply, even in the absence of a specific statement, that such names are exempt from the relevant protective laws and regulations and therefore free for general use.

Cover design: WMX Design, Heidelberg

Printed on acid-free paper

9 8 7 6 5 4 3 2 1

springer.com

Preface

Among renewable sources wind power systems have developed to prominent suppliers of electrical energy. Since the 1980s they have seen an exponential increase, both in unit power ratings and overall capacity. While most of the systems are found on dry land, preferably in coastal regions, off-shore wind parks are expected to add significantly to wind energy conversion in the future.

The theory of modern wind turbines has not been established before the 20th century. Currently wind turbines with three blades and horizontal shaft prevail. The driven electric generators are of the asynchronous or synchronous type, with or without interposed gearbox. Modern systems are designed for variable speed operation which make power electronic devices play an important part in wind energy conversion. Manufacturing has reached the state of a high-tech industry.

Countries prominent for the amount of installed wind turbine systems feeding into the grid are in Europe Denmark, Germany and Spain. Outside Europe it is the United States of America and India who stand out with large rates of increase. The market and the degree of contribution to the energy consumption in a country has been strongly influenced by National support schemes, such as guaranteed feed-in tariffs or tax credits.

Due to the personal background of the author, the view is mainly directed on Europe, and many examples are taken from the German scene. However, the situation in other continents, especially North America and Asia is also considered.

This book was written from the standpoint of electrical engineering. It is meant to provide basic knowledge on wind energy systems for graduate students of technical disciplines, and for engineers who seek overview information apart from their own special field. The intention is to convey the properties and performance of wind rotors and of the electrical components both to the electrical and the mechanical engineer.

The author wishes to extend special thanks to his colleagues Prof. Dr. R. Gasch and Prof. Dr. J. Twele to whom he owes valuable communication during a year-long cooperation at TU Berlin in common projects pertaining to wind energy. Also, thanks are due to Springer-Verlag for their care during the book production.

Berlin, Germany
M. Stiebler

Contents

1 Role of Wind as a Renewable Energy ... 1
  1.1 Renewable Energies and Their Application ... 1
    1.1.1 Sources of Renewable Energy ... 1
    1.1.2 Sources of Electrical Energy Production ... 2
    1.1.3 Wind and Solar Energy ... 3
  1.2 Wind Energy Contribution to Electrical Supply ... 3
    1.2.1 Installed Power ... 3
    1.2.2 Technical Standardization and Local issues ... 7
    1.2.3 Governmental Regulations ... 8

2 Wind Turbines ... 11
  2.1 General ... 11
  2.2 Basics of Wind Energy Conversion ... 11
    2.2.1 Power Conversion and Power Coefficient ... 11
    2.2.2 Forces and Torque ... 14
  2.3 Wind Regime and Utilization ... 17
    2.3.1 Wind Velocity Distribution ... 17
    2.3.2 Power Distribution and Energy ... 18
    2.3.3 Power and Torque Characteristics ... 19
  2.4 Power Characteristics and Energy Yield ... 20
    2.4.1 Control and Power Limitation ... 20
    2.4.2 Wind Classes ... 22
    2.4.3 System Power Characteristics ... 23
    2.4.4 Annual Energy Yield ... 26

3 Generators ... 29
  3.1 General ... 29
  3.2 Asynchronous Machines ... 30
    3.2.1 Principles of Operation ... 30
    3.2.2 Performance Equations and Equivalent Circuits ... 31

viii Contents

		3.2.3	Reactive Power Compensation	40
		3.2.4	Self-Excited Operation	41
	3.3	Synchronous Machines		43
		3.3.1	Principles of Operation	43
		3.3.2	Performance Equations and Equivalent Circuits	44
		3.3.3	Unconventional Machine Types	48
	3.4	Generator Comparison		52

4 Electrical Equipment 55
  4.1 General 55
  4.2 Conventional Electrical Equipment 55
  4.3 Power Electronic Converters 55
    4.3.1 General 55
    4.3.2 External-Commutated Inverters 57
    4.3.3 Self-Commutated Inverters 61
    4.3.4 Converters with Intermediate Circuits 67
    4.3.5 D.c./d.c. Choppers 67
    4.3.6 A.c. Power Controllers 70
  4.4 Energy Storage Devices 71
    4.4.1 General 71
    4.4.2 Electrochemical Energy Storage 71
    4.4.3 Electrical Energy Storage 75
    4.4.4 Mechanical Energy Storage 76

5 Wind Energy Systems 81
  5.1 General 81
  5.2 Systems Overview 81
    5.2.1 General 81
    5.2.2 Systems Feeding into the Grid 82
    5.2.3 Systems for Island Supply 83
    5.2.4 Wind Pumping Systems with Electrical Power Transmission 84
  5.3 Systems for Feeding into the Grid 84
    5.3.1 General 84
    5.3.2 Induction Generators for Direct Grid Coupling ... 85
    5.3.3 Asynchronous Generators in Static Cascades 86
    5.3.4 Synchronous Generators 95
    5.3.5 Examples of Commercial Systems 103
  5.4 Systems for Island Operation 103
    5.4.1 Systems in Combined Generation 103
    5.4.2 Stand-Alone Systems 106

6 Performance and Operation Management 115
  6.1 General 115
  6.2 System Component Models 115
    6.2.1 Model Representation 115

Contents ix

	6.2.2	Asynchronous Machine Models	120
	6.2.3	Synchronous Machine Models	127
	6.2.4	Converter Modeling	133
	6.2.5	Modeling the Drive Train	135
6.3	System Control		138
	6.3.1	General	138
	6.3.2	Control of Systems Feeding into the Grid	139
6.4	Basics of Operation Management		143
	6.4.1	General	143
	6.4.2	States of Operation	143
	6.4.3	Grid Fault Reaction	144

7 Grid Integration and Power Quality 147

7.1	Basics of Grid Connection		147
	7.1.1	General	147
	7.1.2	Permissible Power Ratings for Grid Connection	147
	7.1.3	Power Variation and Grid Reaction	150
7.2	Standard Requirements		152
	7.2.1	Safety-Relevant Set Values	152
	7.2.2	Reactive Power Compensation	152
	7.2.3	Lightning Protection	152
7.3	System Operator Regulations		154
	7.3.1	General	154
	7.3.2	Active Power and Frequency	155
	7.3.3	Reactive Power and Voltage	156
	7.3.4	Short-Circuit and Fault Ride-Through	158
7.4	Power Quality		158
	7.4.1	Harmonics	158
	7.4.2	Voltage Deviations and Flicker	159
	7.4.3	Audio Frequency Transmission Compatibility	165
7.5	Noise Emission		166
	7.5.1	General	166
	7.5.2	Sound Emission by WES	167

8 Future of Wind Energy 171

8.1	Off-Shore Wind Energy Generation		171
	8.1.1	General	171
	8.1.2	Foundation	171
	8.1.3	Connection	172
	8.1.4	Specific Issues and Concerns	174
8.2	Power Integration and Outlook		177
	8.2.1	Wind Energy in Power Generation Mix	177
	8.2.2	Integration in Supranational Grids	177
	8.2.3	Outlook on 2020	177

Annex A – List of Symbols 179

Annex B – List of Abbreviations 183

Bibliography 185

Index 189

List of Figures

1.1	World net generation of electricity (2004) (**a**) per world part and source, in TWh, (**b**) visualization of components	2
1.2	Wind energy installed power ratings and annual output (2002). (**a**) installed capacity, in MW; (**b**) electric output; in GWh	4
1.3	Installed capacity as total and new in 2006 in top 10 countries	5
1.4	Concentration of installed wind power in the EU	6
1.5	Average installed power per unit in Germany, in kW	6
1.6	Schemes of renewable energy support in the EU 25	9
2.1	Idealized fluid model for a wind rotor (Betz)	12
2.2	Typical power coefficients of different rotor types over tip-speed ratio	13
2.3	Typical torque coefficients of different rotor with hotizontal shaft	14
2.4	Coefficients $c_A(\alpha)$ of lift and $c_W(\alpha)$ of drag over blade angle of specific profiles	15
2.5	Wind speeds and forces acting on the blade	15
2.6	Curves of power coefficient $c_p(\lambda)$ and torque coeffient $c_T(\lambda)$ of a three-blade rotor	16
2.7	Curve of drag (drag) coefficient $c_S(\lambda)$	16
2.8	Representation of wind velocity distribution. (**a**) example histogram; (**b**) approximation by Weibull-functions (Raileigh-function for $k = 2$)	18
2.9	Histograms of wind velocity distribution and normalized energy yield	18
2.10	Power and torque characteristics vs. rotational speed ($v_N = 12\,\mathrm{m/s}$)	19
2.11	Power, torque and drag coefficients over tip speed ratio with pitch angle as parameter	21
2.12	Sketch of a blade with laminar and turbulent air flow	21
2.13	Illustration of stall, active-stall and pitch effects	22
2.14	Typical power curves for pitch-controlled and stall-controlled systems	23
2.15	Sketch of a the measuring setup	24
2.16	Power curve of a system specified for 1800 kW	24
2.17	Rotor diameters and rotation speeds of systems of 850 kW and above	25

xi

2.18 Specific power of systems > 850 kW and reference energy yield of systems 2000 kW 25

2.19 Full-load equivalent annual energy yield over ratio rated/average wind speed 26

2.20 Energy yield kWh/m^2 per month as recorded in Germany and Austria 26

3.1 Diagram of three-phase induction machine with wound rotor 30

3.2 Asynchronous machine T-model circuit 31

3.3 T-model for asynchronous machine connected to the grid 32

3.4 T-model variants containing loss resistors (a) conventional iron loss resistor R_{Fe}; (b) resistor R_p representing constant losses 33

3.5 Principal characteristics of current and torque at constant flux linkage 34

3.6 Performance in steady-state (a) current locus (Ossanna's circle diagramme); (b) vector diagramme (Generator operation) 35

3.7 Operation at rated voltage (a) Performance characteristics; (b) Current and torque vs. speed and slip 36

3.8 Sankey diagrams of induction machines (a) motor operation; (b) generator operation 37

3.9 Flux and inductance under main field saturation 39

3.10 Compensation device concepts (a) basic circuits; (b) admittance characteristics 40

3.11 Induction machine capacitive self-excitation 41

3.12 Self-excited induction generator with passive load (a) Circuit diagram; (b) Equivalent circuit with ohmic load only 42

3.13 Load characteristics of a laboratory setup with SEIG and ohmic load (a) curves at $n = $ const, capacitance C as parameter; (b) curves at C = const, speed n as parameter 43

3.14 Diagram of three-phase synchronous machine with separate excitation 43

3.15 Equivalent circuits of the turbo-type synchronous machine (a) with voltage source $\underline{U}_p$; (b) with current source $\underline{I}_f$ 44

3.16 Steady-state characteristics of a synchronous machine in grid-operation (a) Current locus diagram; (b) vector diagram 45

3.17 Torque characteristic in grid operation. (a) Turbo-type machine; (b) Salient polemachine, $X_q < X_d$ 46

3.18 Operation with passive R, L load (a) Equivalent circuit for turbo-type machine; (b) load curves vs. normalized frequency 47

3.19 Output power characteristic for ohmic load 47

3.20 Principal types of axial field machines 49

3.21 Flux path sketch in axial field machines 49

3.22 Small axial generator with air-gap ring winding 50

3.23 Sketch of modular axial field machine 50

3.24 Principle of transversal flux machines (a) single-sided component of polyphase machine; (b) Double-sided with intermediate rotor 51

4.1	Power electronics applications	56
4.2	Six-pulse bridge (B6) thyristor rectifier circuit	57
4.3	Simplified waveforms of B6 circuit under load (a) Rectifier operation, uncontrolled ($\alpha = 0$) or diode; (b) Rectifier operation, controlled ($0 < \alpha < \pi/2$), example	59
4.4	Waveforms of M3 circuit under load, (a) rectifier, $\alpha = 22°$, (b) a.c inverter, $\alpha = 142°$	59
4.5	Circle diagram showing the relation between d.c voltage and control-reactive power (u_0 values indicate initial commutation overlap angles)	60
4.6	Reactive power inverter with inductive storage element (a) circuit; (b) voltage and currents (example)	61
4.7	Full-wave bridge voltage-source inverter	61
4.8	Voltage waveforms in six-step operation	62
4.9	PWM bridge (B6) inverter circuit	63
4.10	Voltage waveforms in PWM operation with sinusoidal modulation	63
4.11	Model of generator, coupling inductor and inverter	64
4.12	Phasor diagram of a.c. fundamentals in different cases of operation	65
4.13	Power locus diagram in consumer (motor) coordinates, for lossless inductor	66
4.14	Reactive power inverter with capacitive storage element, in square-wave operation (a) circuit; (b) voltage and currents (example)	66
4.15	Converter schemes with intermediate circuits. (a) voltage-source inverter (VSI); (b) current-source inverter (CSI)	67
4.16	D.c./d.c. chopper with inductor storage element. (a) Step-down (buck) converter; (b) Step-up (boost) converter	68
4.17	Step-up converter, continuous and discontinuous conduction. (a) bondary case of cont./discont. conduction; (b) boundary curves; $I_{ref} = U_o T/L$	69
4.18	Step-up converter, control characteristics	69
4.19	Three-phase power controlers (a) Circuit with ohmic-onductive load; (b) Control characteristic of r.m.s. current	70
4.20	Ragone diagram of energy storage devices	72
4.21	Cell voltage U over degree of charge p	72
4.22	Characteristic of a conventional battery (a) Available capacity; (b) Battery voltage during discharging	73
4.23	Charging method using I/U control (example)	73
4.24	Battery equivalent circuit	74
4.25	Concept of a CAES plant (Huntorf/Germany)	78
5.1	Typical concepts for generating electrical power. – using induction generator: (a) direct coupling, (b) fully-fed, (c) doubly fed, – using synchronous generator, fully fed; (d) electrical excitation. (e) PM excitation	82
5.2	Common concepts of systems feeding into the grid (Legend see text)	83

5.3	Shares of type groups installed in 2004 in Germany	85
5.4	Normalized power and torque characteristics of a wind turbine with fixed blades	86
5.5	Power curve of a stall turbine system	87
5.6	Principle diagram of a constant speed system	88
5.7	Operation and control of a wind energy system. Sketch of data acquisition (*left*) and block diagram of operation management (*right*)	89
5.8	L-model for asynchronous wound-rotor machines	89
5.9	Static Kramer system, basic circuit diagram	90
5.10	Steady state operation of the wound rotor induction machine (**a**) complex current locus for constant rotor voltage ($s_0 = \pm 0,1$ as example); (**b**) phasor diagram for operation as a generator	91
5.11	Visualization of Kramer cascade operation (**a**) Diagram for slip power recovery (losses neglected); (**b**) Sankey diagrams; left: motor operation ($s > 0$); right: generator operation ($s < 0$)	92
5.12	Torque and power characteristics of the cascade system (**a**) Torque characteristics; (**b**) Power characteristics, example for $k_2 = 0,1$	93
5.13	Doubly-fed induction generator with rotor-side converter	93
5.14	Steady-state performance of a doubly-fed induction machine (**a**) current loci; (**b**) variable speed operation (legend see text)	94
5.15	Synchronous generator with full-load converter (legend see text)	95
5.16	Synchronous generator with slipringless excitation	96
5.17	Demagnetization curves of different magnetic materials	98
5.18	Demagnetization curves of aNdFeB magnets material [VAC]	99
5.19	PMSM characteristics (**a**) Magnetization; (**b**) field line pattern	99
5.20	Concept of an island grid with combined generation and storage	104
5.21	Variants of frequency control	105
5.22	Concept of an autonomous island system with renewable energy generation	106
5.23	Characteristics of a battery loader (example)	107
5.24	Sketch of furling action	108
5.25	Turbine diameters and maximum speed of commercially available small WES	108
5.26	Typical inverter circuits for small ratings	109
5.27	Circuit of a system with step-up inverter	110
5.28	Performance of a system with step-up inverter and battery storage (example) (**a–b**) voltages, currents; (**c**) duty ratio; (**d**) input power	110
5.29	Stand-alone system with synchronous generator and battery storage	111
5.30	Self-excited induction generator with phase control	112
5.31	Stand-alone system with cage induction machine and battery storage	112
6.1	Three-phase machine model and coordinates (**a**) three-phase windings in stator and rotor; (**b**) different frames	120
6.2	Induction machine model in Clarke components	121
6.3	Dynamic induction machine model	123

List of Figures xv

6.4 Rotor model of induction machine (**a**) Space-vector diagram;
(**b**) block diagram 125
6.5 Synchronous machine model (**a**) Salient pole three-phase machine;
(**b**) Transformed windings arrangement 128
6.6 Model of the synchronous machine with five windings (**a**) windings
arrangement in d,q components; (**b**) equivalent circuit model 132
6.7 Four-pole equivalent circuit model in hybrid form 134
6.8 Model of intermediate circuit voltage source converter 135
6.9 Three-mass drive train model of a wind system T wind turbine;
B gear box; G generator 136
6.10 Resonance curves of displacement angle Φ (*left*) and shaft torque
T_{sh} (*right*) Damping coefficients $d = 0,05; 0, 1; 0, 2; 0, 4; 0, 8; 1, 6$. 137
6.11 Analogy of PV- and wind energy systems 139
6.12 Control scheme of a system for constant speed operation 140
6.13 Control scheme of a system for variable speed operation 141
6.14 Control scheme of a WES with doubly-fed induction generator 141
6.15 Waveforms of the WES quantities during an example wind regime . . 142

7.1 Model of short-circuit impedance between generator and grid and
voltage variation 149
7.2 Example of measured power variations 150
7.3 Periodic and non-periodic voltage distortions. (1) oscillatory
transient; (2) voltage sag; (3) voltage swell; (4) momentary
interruption; (5) voltage flicker; (6) harmonic distortion; (7) voltage
with interharmonics; (8) voltage with notches 151
7.4 Reactive power distribution example of a 50 MW wind park 153
7.5 Lightning protection concept 153
7.6 Quadrant definition in consumer system 155
7.7 Required active power capability of WES supplying a high-voltage
grid 156
7.8 Reactive power supply requirements 157
7.9 Power factor assigned to grid voltage area 157
7.10 Limiting voltage/time area excluding automatic tripping of wind
parks 158
7.11 Limiting curve of voltage variations per minute 160
7.12 Block diagram of a flickermeter 161
7.13 Typical signal waveforms in a flickermeter 161
7.14 Example from an arc furnace of a cumulative flicker power curve
to determine P_{lt} 163
7.15 Flicker-relevant voltage drop on a short-circuit impedance 165
7.16 Model of capacitive compensator with series inductor 166
7.17 Relative impedance of series resonant circuit 167
7.18 Measured sound vs. wind velocity 168
7.19 Sound power levels of wind energy systems 169
7.20 Curves of equal sound pressure level in the vicinity of a WES 169

8.1	Foundation structures	172
8.2	Cable Π-model and limiting length	173
8.3	Principal comparison of HVAC and HVDC connection cost	174
8.4	Concepts of electrical connection of a wind farm 100 MW. (a) AC connection; (b) DC connection	175
8.5	EU electricity projection by 2020	178

List of Tables

1.1	Sources of renewable energy	2
1.2	Installed wind capacity in Europe, in MW (2006)	5
2.1	IEC type classes	22
4.1	Properties of accumulators	75
4.2	Main properties of CAES plant Huntorf	78
5.1	Properties of selected wind energy systems	100
5.1	(continued)	101
5.1	(continued)	102
5.2	Comparison of inverter circuits in Fig. 5.26	109
6.1	Transformation matrices in the power-variant form I	119
6.2	Transformation matrices in the power variant form II	119
6.3	Properties of preferred transformations	124
7.1	Permissible harmonic currents at connecting point	159
7.2	Characteristics of selected systems (see Table 5.1) as determined from tests	164
7.3	Noise limits established by TA Laerm	169
8.1	Basic comparsion of AC and DC connections	176
8.2	HVDC design features as compared with the classical concept	176

xvii

Chapter 1
Role of Wind as a Renewable Energy

1.1 Renewable Energies and Their Application

1.1.1 Sources of Renewable Energy

It is commonly accepted that the earth's fossil energy resources are limited, and the global oil, gas and coal production will come beyond their peak in the next decades, and price rises will continue. At the same time there is strong political opposition against strengthening nuclear power in many parts of the world. In this scenario renewable energies will have to contribute more and more to the world's ever rising need of energy in the future [Bul01]. Renewables are climate-friendly forms of energy, due to the absence of emissions detrimental to the environment. The savings especially in carbon-dioxide and sulphur dioxide emissions are a significant advantage over fossil power stations. Hence a main role is assigned to renewable energy in the proclaimed fight against Climate Change.

The major source of renewable energies is the sun, with some forms also attributed to the earth and the moon. Table 1.1 lists the primary sources, the natural ways of conversion and the technically used conversion methods. Notable for their contribution to the current energy demand are water, wind, solar energy and biomass. Utilization of renewables is mostly with conversion into electrical energy.

While water power has been used in electrical power stations and pumped storage systems since many decades, the use of wind power conversion in larger ratings has begun only in the 1980s. Backed by intense technical development, unit ratings have grown fast into the MW range, and wind parks were erected in large numbers with considerable increase rates.

Solar energy is applied both by direct conversion in photovoltaic generators, and via thermal collectors and steam production. The energy from renewable sources is partly already competitive in price, and partly supported by state legislative to promote their share in the market. Wind energy systems are about to reach the competitiveness before long, while photovoltaic energy production is still expensive and will require further support on their way to market relevance.

M. Stiebler, *Wind Energy Systems for Electric Power Generation*. Green Energy and Technology, © Springer-Verlag Berlin Heidelberg 2008

Table 1.1 Sources of renewable energy

Primary source	Medium	Natural conversion	Technical conversion
Sun	Water	Evaporation, precipitation, melting	Water power plants
	Wind	Atmospheric airflow	Wind energy conversion
		Wave movement	Wave power plant
	Solar energy	Ocean current	Ocean power plant
		Heating earth surface and atmosphere	Thermal power units, heat pumps
		Solar radiation	Heliothermal conversion, Photovoltaic conversion
	Biomass	Biomass production	Co-generation plants
Earth	Isotop decay	Geothermal heat	Co-generation plants
Moon	Gravitation	Tides	Tide power plants

1.1.2 Sources of Electrical Energy Production

Looking at the sources of energy currently applied world-wide for generating electricity, it can be seen that fossil fuels in form of coal, oil and natural gas prevail (65%). Nuclear energy (16%) and hydro energy (17%) follow with almost same percentage. In the representation of Fig. 1.1 the part indicated as Renewables (2%)

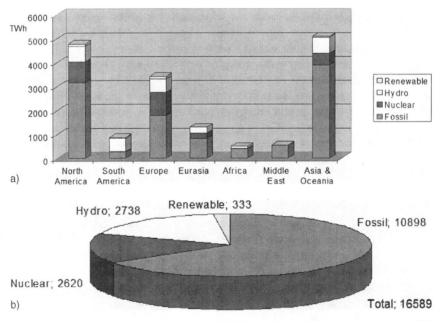

Fig. 1.1 World net generation of electricity (2004) (**a**) per world part and source, in TWh, (**b**) visualization of components

covers mainly wind, biomass and solar energy, but is shown separately from hydro energy, while it is understood that the latter is the energy form contributing the best part among to the useful renewables [WEC04].

1.1.3 Wind and Solar Energy

It may be worth noting that the per-area power densities offered by kinetic wind energy and solar radiation are in the same order of magnitude, when considering exploitable values in specific regions. As an example, the wind exerts at 20 m/s on a vertical plane $1,04 \, \text{kW/m}^2$, while the solar radiant flux density on a horizontal plane is e.g. (on 21st June at noon, latitude 50 north) $1,05 \, \text{kW/m}^2$. Both forms of renewable energy are characterized by non-steady regime. The efficiency of power conversion is $40 \ldots 50\%$ for wind systems, and $12 \ldots 18\%$ for photovoltaic generation with current commercially available silicon cells.

The order of the annual reference energy yield per swept area is roughly 800 $\ldots 1000 \, \text{kWh/(m}^2\text{a)}$ for European in-land wind parks (at 5,5 m/s average wind velocity, 1700 h annual full-load hours and 0,45 best point efficiency). For photovoltaic systems $700 \ldots 900 \, \text{kWh/kW}_p$ are relevant values, e.g. for Northern Germany. Assuming a specific PV-generator area of $5 \, \text{m}^2/\text{kW}_p$, this corresponds to $160 \, \text{kWh/(m}^2\text{a)}$ yield per PV converter area. In these values the system losses up to the generator or converter terminal output, respectively, are taken into account. Admittedly the reported magnitudes are relatively low when compared with coal, oil and gas fueled power stations, while obviously a fair per area comparison cannot be made.

1.2 Wind Energy Contribution to Electrical Supply

1.2.1 Installed Power

By the end of 2002 the rated installed power in wind farms was roughly 32 GW worldwide. Statistical values of regional distribution are given in Fig. 1.2 [eia]. It is obvious that Europe has been the leader in wind power utilization, contributing 76% of the total power. North America which follows by 16% has since increased its percentage considerably.

As by 2006 roughly 65 GW of rated power were installed in wind farms worldwide, of which more than 47 GW located in the countries of the European Community, and more than 11 GW in the United States. The Wind capacity installed in 2007 was almost as high as 20 GW [GWEC, 08 update], so that for the the end of 2007 a cumulative 94 GW were reported, where the EU stands for 56,5 GW and the USA contribute 16,8 GW. The exorbitant progress since the 1980s was accompanied by a significant decrease in cost per kWh, due to technical development

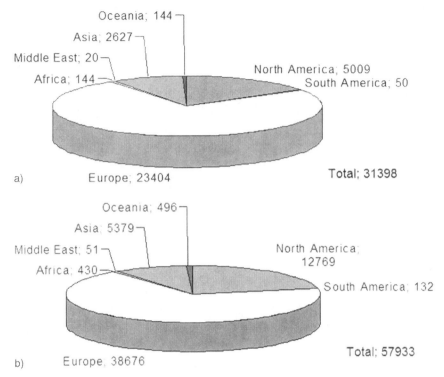

Fig. 1.2 Wind energy installed power ratings and annual output (2002). (**a**) installed capacity, in MW; (**b**) electric output; in GWh

and the economies of large-scale generation. Sources of progress were technological advances in structural analysis and design, sophistication in blade design and manufacture, and the utilization of power electronics and variable-speed operation [Ram07]. Hence wind electric conversion is the fastest growing "green" technology.

A survey of the Top 10 countries regarding installed wind energy conversion capacity is shown in Fig. 1.3 [GWEC07]. Among the countries other than Europe, USA, India and China are notable for using wind energy with increasing capacities. Wind reports regarding the 27 member countries of the International Energy Agency (IEA) per 2006 are collected in [IEA06].

Focussing on the situation in Europe, Table 1.2 gives a list of rated power in MW, installed during 2006 and total at the end of 2006. Germany is the leading country by installed wind power, with Spain following, while Denmark is foremost in capacity per capita. Reported in the table are also the total capacities for the European Union (EU) of 15 countries, of 25 countries and for total Europe [ewea].

Within the EU the concentration of installed wind power in 10 countries is also reflected in a European Wind Integration Study (EWIS), Final Report of 2007-01 [EWIS07], see Fig. 1.4.

In a special view on Germany, the percentage of electrical energy supplied by renewables has reached 14.8% of the national consumption, of which almost half

1.2 Wind Energy Contribution to Electrical Supply

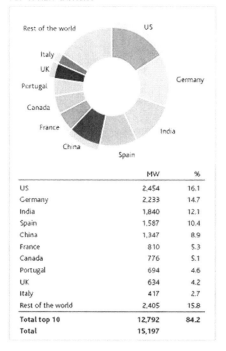

Fig. 1.3 Installed capacity as total and new in 2006 in top 10 countries

Table 1.2 Installed wind capacity in Europe, in MW (2006)

Countries	Installed in 2006	Total end 2006
Germany	2233	20622
Spain	1587	11615
Denmark	12	3136
Italy	417	2123
UK	634	1963
Portugal	694	1716
France	810	1567
Netherlands	356	1560
EU 15	**7404**	**47644**
EU 25	**7588**	**48027**
Total Europe	**7708**	**48545**

(7.2%) was due to wind power. values of 2007. In some coastal regions this has led to a situation where in times of low load the supplied wind power can be larger than the consumption. The issue is to transfer the energy to regions of conurbation and centers of industry. It is realized that a reinforcement of high voltage lines is necessary. Expected is the need for 1050 km of new high-voltage lines until 2020.

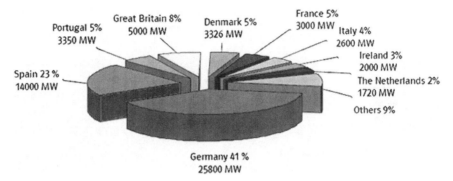

Fig. 1.4 Concentration of installed wind power in the EU

Due to the meteorological variations in wind availability, energy storage is also an important issue of rising interest.

Another aspect is that the utilities have begun to establish additional requirements, in order to have the wind park operators contribute to the supply of reactive power and provide ride-through capabilities in case of grid failures, similar to what is expected from conventional power stations. This poses a challenge to the manufacturing industry.

The development of wind turbines is reflected in the increase of power per installed unit. Figure 1.5 shows the values for Germany from 1987 to 2007 [End08]. It is seen that the average generator power has reached 1800 kW. Units of 2, 3 ... 3 MW are commercially available; and in course of installing future offshore windparks

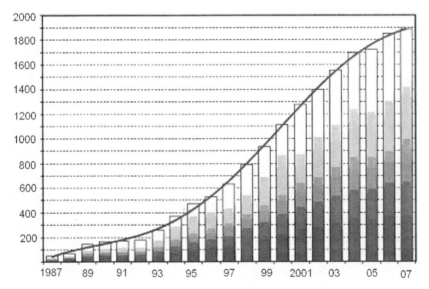

Fig. 1.5 Average installed power per unit in Germany, in kW

1.2 Wind Energy Contribution to Electrical Supply

there will be further increase of average rated power in the Megawatt class. At present units of up to 5 MW are in operation as pilot systems.

1.2.2 Technical Standardization and Local issues

Rapid technical and economical progress has led to the present situation where wind energy is used on industrial scale and is meanwhile in some countries contributing a considerable share to national energy consumption. While the profusion of wind parks has been supported by National legislature in several regions, wind energy is expected to become competitive in the near future.

The state of the art is being laid down in technical standards. Internationally relevant documents have been published by the International Electrotechnical Commission (IEC) [IEC61400]. These standards are developed by international groups of experts and revised in appropriate intervals. They provide guidelines for the design, establish safety regulations and address environmental issues such as noise emission and power quality of wind energy systems.

Wind farm installation starts with the assessment of the annual wind regime in the location selected for erection. Cost of technical equipment and maintenance are figures in the consequent economic study, leading to a prospect of revenue within, say, 20 years. The owners of wind parks reflect a wide variety of investor types, comprising local groups of rural land owners, private companies of limited partnership and institutional investors including the regional utilities themselves. It must be noted that in some countries with already high degree of wind power utilization the identification of locations for erecting wind farms and the process of obtaining permission from regional and local bodies has become difficult and time-consuming.

It is almost generally accepted that the utilization of renewable energies is, besides energy saving, the best means to reduce pollution and decelerate climate change. Arguments denouncing wind system technology as inefficient and too costly are heard less in recent years. On the other hand, issues of features detrimental to environment in a broad sense are subjects of public discussion, eventually slowing down the progress of wind utilization:

– Noise emission
  Like any machine WES are sources of noise emission. Aerodynamic noise of the blades prevails over other components. To minimize noise emission manufacturers have made intensive efforts. To protect humans in the vicinity of WES, noise limits have been established (see 7.7). Besides noise in the audible frequencies so-called infra-noise has also been the subject of concern.
– Oscillating shadow
  The oscillating shadow of a WES due to the rotating blades optical can also be a source of optical disturbance for residents ("disco effect"). Depending on local conditions, minimum distances are required, e.g. 6 times the overall height as mandated by a court.

8 1 Role of Wind as a Renewable Energy

– Animals and habitat
It is understood that nature reserves and National parks are not open for installing WES. The main focus is, however, on the danger posed by wind turbines for flying birds. Many investigations have been made; roughly the results is that birds are less endangered than was feared by nature preserving organizations. More detailed studies are still underway.
– Appearance in the landscape
Undoubtedly WES have an influence on the landscape phenotype. In regions where tourists are of economical importance it has been argued that visitors may be repelled by the negative impact of extensive wind mast installations in the landscape.

In effect, in some inland regions of countries like Germany it is has become hard to identify further locations for wind parks. Offshore wind parks are expected to ease the situation, however they give rise to other environmental concerns (see 8.1).

1.2.3 Governmental Regulations

In Europe there has been a strong policy support both at EU and national level. The EU Renewable Directive (77/2001/EC) is in place since 2001. The aim is to increase the share of energy produced from renewable energy sources to 21% by 2010. According to the directive, each member state is obliged to generate a specific percentage of its electrical energy from renewables. It is up to the member states to promote renewables by appropriate measures. The EU member states apply different schemes to promote generation and consumption. The methods vary from guaranteed feed-in tariffs, fixed premium, green certificate systems to tendering procedures, complemented by tax incentives, environmental taxes, contribution programs and voluntary agreements.

An overview of schemes supporting renewable electrical energy in the EU 25 is given in Fig. 1.6 [ISI06]. In the USA renewable energies are promoted by granting Production Tax Credits (PTC), the relevant schedules decided by federal legislation [AWEA].

Where feed-in tariffs exist (e.g. in Germany), utilities are obliged to enable wind parks to connect to the grid, and to pay a minimum fixed price for the supplied renewable energy. In specific regulations the price is guaranteed over a certain time, reductions being applied normally with increasing age of the systems. The additional cost above conventional generation is passed on to the consumers. Feed-in tariffs have proven to offer investment security and contain incentives for technological progress. A variant of feed-in tariffs is a fixed-premium scheme (e.g. Denmark, Spain), where a bonus is paid to the producers above the normal electricity price.

In the Green Certificate scheme (e.g. Belgium, Italy, Poland, UK and Sweden) the producers sell electrical energy at market prices, within a quota set be the government. The consumers have to purchase from the producers. Penalties are established for non-compliance. A secondary market for green certificates comes into existence

1.2 Wind Energy Contribution to Electrical Supply

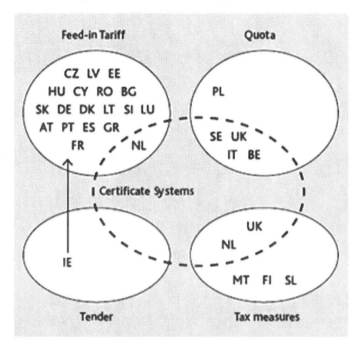

Fig. 1.6 Schemes of renewable energy support in the EU 25

where producers compete with one another. This method gives less long-time investment security than the feed-in tariff does.

In the tendering system (Ireland, formerly France) the state places tenders for the supply of renewable energy, and the electricity is sold at the resulting price. The additional cost is passed on to the consumers. In spite of theoretical benefits the scheme tends to unstable conditions for investments.

The European Commission, in a report of December 2005, has assessed the various support schemes, but has refrained from proposing a harmonization at this stage. Nevertheless, it turned out that the feed-in tariffs have currently the best performance in the wind energy application.

Looking specifically at Germany, the "Erneuerbare Energien Gesetz" (Renewable Energy Act) has succeeded an earlier "Strom Einspeise Gesetz" (Electricity Supply Act) of 1999. The current law [BRD04], with an amendment in force January 2009, guarantees the owners of wind parks the right to feed any amount of electrical energy at any time into the public grid at fixed price, provided an agreed level of power quality is kept. Detailed regulations apply to the different forms of renewable energy including wind, describing system reference power values and kWh-prices which decrease with the years elapsing after commissioning. The difference between the fixed price and the average market price of energy from conventional sources is charged on the consumers by an addition to the basic price. The law was the result of long political discussion but is currently the accepted basis for further deliberation in planning the future mix of sources for National energy production.

Chapter 2
Wind Turbines

2.1 General

The history of windmills goes back more than 2000 years. They have been used predominantly for grinding cereals and for pumping water. Important examples of more recent times are the Dutch Windmills which appeared in different variants and were erected in large numbers in the 17th and 18th century in Europe. Another memorable development of the 19th century was the Western Mill, found in rural areas especially in the USA up to the present day. Modern constructions of wind energy converters were developed in the 1920s, but it was not before the 1980s that they found professional interest as a prominent application of renewable energies.

From the standpoint of fluid engineering we have to distinguish wind energy converters with drag force rotor and with buoyant force rotor. While drag force rotors utilize directly the wind pressure and attain only low efficiencies in the order of $0, 1 \ldots 0, 2$, buoyant force rotors develop considerably higher values, inherently limited at approximately 0.59. A suitable theory was published not before the start of the 19th century (Joukowski 1907).

Modern windturbines are mostly constructed as fast running machines with horizontal shaft, upwind arrangement and preferably 3 rotor blades. The machine ratings have steadily increased so that the average installed power per unit is currently above 1.700 kW. For offshore-wind-parks ratings of up to 6.000 kW are in pilot stage.

A selection of relevant book literature on wind turbines is in [Hau06, Gas02, Gas07].

2.2 Basics of Wind Energy Conversion

2.2.1 Power Conversion and Power Coefficient

From the expression for kinetic energy in flowing air follows the power contained in the wind passing an area A with the wind velocity v_1:

$$P_w = \frac{\rho}{2} A v_1^3 \tag{2.1}$$

M. Stiebler, *Wind Energy Systems for Electric Power Generation*. Green Energy and Technology, © Springer-Verlag Berlin Heidelberg 2008

Here ρ is the specific air mass which depends on air pressure and moisture; for practical calculations it may be assumed $\rho \approx 1.2\,\text{kg/m}^3$. The air streams in axial direction through thea wind turbine, of which A is the circular swept area. The useful mechanical power obtained is expressed by means of the power coefficient c_p:

$$P = c_p \frac{\rho}{2} A v_1^3 \qquad (2.2)$$

In case of homogenous air flow the wind velocity, whose value before the turbine plane is v_1, suffers a retardation due to the power conversion to a speed v_3 well behind the wind turbine, see Fig. 2.1. Simplified theory claims that in the plane of the moving blades the velocity is of average value $v_2 = (v_1 + v_3)/2$. On this basis Betz [Bet26] has shown by a simple extremum calculation that the maximum useful power is obtained for $v_3/v_1 = 1/3$; where the power coefficient becomes $c_p = 16/27 \approx 0{,}59$. In reality wind turbines display maximum values $c_{p,\text{max}} = 0{,}4 \ldots 0{,}5$ due to losses (profile loss, tip loss and loss due to wake rotation). In order to determine the mechanical power available for the load machine (electrical generator, pump) the expression (2.2) has to be multiplied with the efficiency of the drive train, taking losses in bearings, couplings and gear boxes into account.

An important parameter of wind rotors is the tip-speed ratio λ which is the ratio of the circumferential velocity of the blade tips and the wind speed:

$$\lambda = u/v_1 = \frac{D}{2} \cdot \frac{\Omega}{v_1} \qquad (2.3)$$

Here D is the outer turbine diameter and Ω is the angular rotor speed. Note that the rotational speed n (conventionally given in min^{-1}) is connected with Ω (in s^{-1}) by $\Omega = 2\pi n/60$.

Considering that in the rotating mechanical system the power is the product of torque T and angular speed Ω ($P = T \cdot \Omega$), the torque coefficient c_T can be derived from the power coefficient:

$$c_T(\lambda) = \frac{c_p(\lambda)}{\lambda} \qquad (2.4)$$

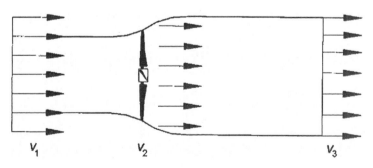

Fig. 2.1 Idealized fluid model for a wind rotor (Betz)

2.2 Basics of Wind Energy Conversion

The torque connected with the power according to (1.2) is then

$$T = c_T \frac{D}{2} \frac{\rho}{2} A v_1^2 \qquad (2.5)$$

Note that the torque varies with the square (v_1^2) and the power varies with the third power (v_1^3) of the wind speed.

Figure 2.2 shows typical characteristics $c_p(\lambda)$ for different types of rotor. Besides the constant maximum value according to Betz the figure indicates a revised curve c_p(Schmitz) which takes the downstream deviation from axial air flow direction into account. The difference is notable in the region of lower tip speed ratios, as calculated by Schmitz and, before, Glauert. Together with Fig. 2.3 indicating associated characteristics $c_T(\lambda)$, the current preference for three-blade rotors with horizontal shaft is understood. The so-called fast-running turbines with 3, 2 or one blades display the larger values of c_p, while the curves c_T indicate the poor starting torque capability of the fast-running types. Since one and two blade rotors are also problematic with respect to torque variations and noise, the three-blade rotors are currently predominant in all modern wind energy systems. The rotors are normally designed to values $\lambda_A = 5 \ldots 8$.

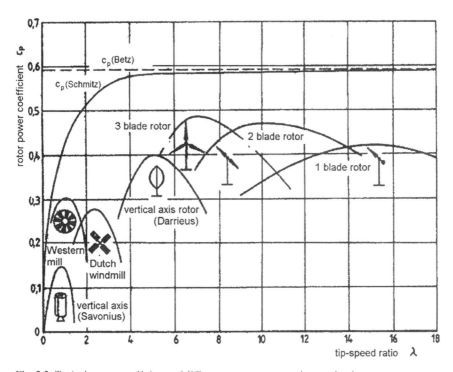

Fig. 2.2 Typical power coefficients of different rotor types over tip-speed ratio

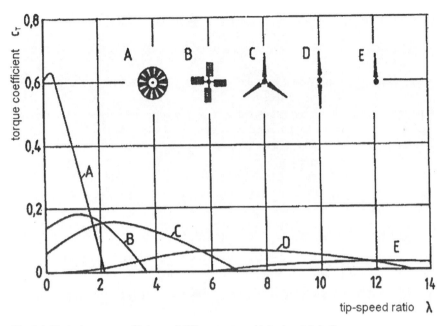

Fig. 2.3 Typical torque coefficients of different rotor with hotizontal shaft

2.2.2 Forces and Torque

The main rotor properties follow from lifting force and drag force of a blade as described by aerofoil theory. Let an aerofoil element of depth t and width b be subjected to a wind speed v_1, see Fig. 2.4. Dependent on the angle of attack α between wind direction and the blade profile cord, the lifting force F_A and drag force F_W are:

$$F_A = c_A(\alpha) \cdot \frac{\rho}{2} v_1^2 \cdot t \cdot b \qquad \text{normal to oncoming flow}$$

$$F_W = c_W(\alpha) \cdot \frac{\rho}{2} v_1^2 \cdot t \cdot b \qquad \text{in direction of oncoming flow} \qquad (2.6)$$

Note that these force components are directed perpendicular and parallel to the oncoming wind, respectively. Coefficients c_A and c_W are characteristic for a given blade profile; they depend on blade angle α. The example in Fig. 2.4 applies to real unsymmetric profiles [Schm56]. For small values of α ($0 \le \alpha \le 10°$) an almost proportional dependence of $c_A = (5,1 \ldots 5,8) \cdot \alpha$ is observed, while c_W is comparatively small in the considered interval of α. The ratio $\varepsilon = c_A/c_W$ is called the glide ratio or lift/drag ratio.

When a wind rotor is rotating at an angular speed Ω, the circumferential speed of each blade at radius r is $u(r) = \Omega \cdot r$. In the rotor plane the wind velocity is v_2 in axial

2.2 Basics of Wind Energy Conversion

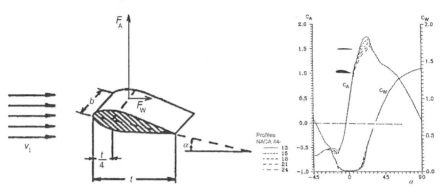

Fig. 2.4 Coefficients $c_A(\alpha)$ of lift and $c_W(\alpha)$ of drag over blade angle of specific profiles

direction, which is, according to Betz theory, 2/3 of the upstream wind velocity v_1. Both components added geometrically result in the speed $c(r)$ which is directed under angle α relative to the rotor plane, see Fig. 2.5. Consequently increments dF_A of lift force and dF_W of drag force are acting on the area increment $(t \cdot dr)$ of the blade. The force can be described by its components dF_t in tangential and dF_a in axial direction:

$$\begin{bmatrix} dF_t \\ dF_a \end{bmatrix} = \frac{\rho}{2} c^2 \, t \, dr \begin{bmatrix} c_A \sin\alpha - c_W \cos\alpha \\ c_A \cos\alpha + c_W \sin\alpha \end{bmatrix} \quad (2.7)$$

Integrating for a given profile, the torque can be obtained from the tangential forces, while the axial forces sum up to the drag force acting axially on the rotor.

At the tip of the blade, $r = R$, the tip speed is $u(R) = \Omega \cdot R$. Note that the wind speed relative to the tip is:

$$c(R) = v_2 \sqrt{1 + \lambda^2}$$

An example of $c_P(\lambda)$ and associated $c_T(\lambda)$ curves for a rotor with fixed blade angle designed for optimum tip-speed ratio $\lambda_A = 6,5$ is shown in Fig. 2.6 [Gas07].

The basic characteristics of a wind rotor follow from the power coefficient $c_P(\lambda)$ and the torque coefficient $c_T(\lambda)$, (see 2.3, 2.4). Further a drag coefficient c_S is

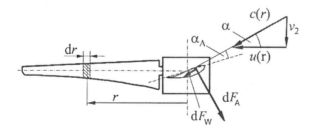

Fig. 2.5 Wind speeds and forces acting on the blade

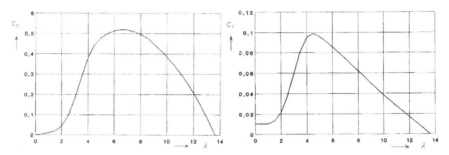

Fig. 2.6 Curves of power coefficient $c_P(\lambda)$ and torque coeffient $c_T(\lambda)$ of a three-blade rotor

defined which allows to calculate the axial drag force, see Fig. 2.7. With rigid blade this coefficient is small for low λ, and attains at no-load speed (large λ) values similar to a circular plane subjected to air flow in normal direction.

Torque T, power P and axial drag force S may be expressed by a set of equations using the reference force F_B which varies with the square of the wind speed and is proportional to the swept rotor area:

$$F_B = \frac{\rho}{2} \frac{D^2 \pi}{4} v_1^2$$

$$T = c_T(\lambda) \frac{D}{2} \cdot F_B \quad ; \quad P = c_P(\lambda) \cdot v_1 \cdot F_B \; ; \; S = c_S(\lambda) \cdot F_B$$

(2.8)

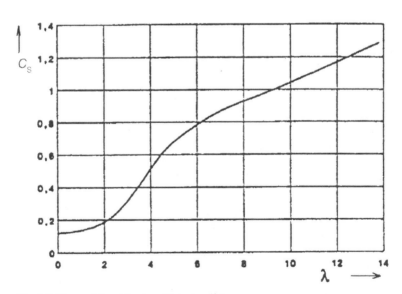

Fig. 2.7 Curve of drag (drag) coefficient $c_S(\lambda)$

2.3 Wind Regime and Utilization

2.3.1 Wind Velocity Distribution

The wind regime is influenced by regional and local effects, and depends on seasonal and short-time variations. In a project to erect a wind energy system a realistic expertise to predict the wind velocity distribution and its average at the relevant place is of foremost importance. The actual wind regime can be determined by a measuring campaign, preferably at the height of the mast. However, data collations show significant variations when considering different years, months and days at a given location.

The wind velocity varies with the height above ground, influenced by the surface roughness. Assuming stable conditions, the dependence of velocity v on height z may be described by a logarithmic profile. The wind speed v_2 at z_2 is calculated from a reference speed v_1 at z_1 by:

$$v_2(z_2) = v_1 \frac{\ln(z_2/z_0)}{\ln(z_1/z_0)} \tag{2.9}$$

where z_0 is the roughness length dependent on the country; conventional parameters are $0.03\,\mathrm{m}$ for farmland, $0.1\,\mathrm{m}$ for heath scattered shrubs and trees, $0.5 \ldots 1.6\,\mathrm{m}$ for forest. Equation (2.9) is used when calculating the reference energy yield in the project stage, see 2.4.4.

When the regime of wind velocity $v(t)$ is known at a specified height above ground, the distribution of power and energy yield can be evaluated by descriptive statistics.

To this end the wind velocities are assigned to k equally distributed classes of width Δv with centre values v_i ($i = 1 \ldots k$). Averages measured during a period of e.g. 10 min are assigned to the k classes, so that each class i is covered by a time interval t_i. The relative frequencies h_i of wind velocity in the period under consideration T of e.g. 1 d are:

$$h_i = \frac{t_i}{T} \quad \mathrm{mit} \quad T = \sum_{i=1}^{k} t_i \tag{2.10a}$$

The frequency distribution is represented in form of a histogram $h_i(v_i)$. Figure 2.8a gives an example with class width $\Delta v = 1\,\mathrm{m/s}$.

It is known that distributions measured in practice may be approximated by a Weibull-function. Distributions obtained in inland Europe follow, with good accuracy, a Weibull-function with form factor $k = 2$, i.e. a Rayleigh-distribution:

$$h_R(\varepsilon) = \frac{\pi}{2} \frac{\Delta v}{v_{av}} \, \varepsilon \cdot \exp\left(-\frac{\pi}{4}\varepsilon^2\right) \quad \text{where} \quad \varepsilon = \frac{v}{v_{av}} \tag{2.10b}$$

Figure 2.8b shows Weibull distributions with k as parameter; note the preference for $k = 2$ in approximation which is the Raileigh distribution widely used in practice.

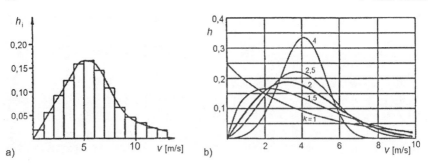

Fig. 2.8 Representation of wind velocity distribution. (**a**) example histogram; (**b**) approximation by Weibull-functions (Raileigh-function for k = 2)

2.3.2 Power Distribution and Energy

A power value $P_i(v_i)$ may be attributed to each class i of the distribution $h_i(v_i)$ according to (2.2). It is advisable to refer P_i to the swept area A, to obtain a specific power value:

$$p_i = \frac{P_i}{A} = c_p \frac{\rho}{2} v_i^3 \qquad (2.11)$$

Note that a power limitation may take place at higher wind velocities. From the specific power follows a normalized energy distribution:

$$e_i = \frac{p_i \cdot h_i}{E} \qquad \text{where} \qquad E = \sum_{i=1}^{k} p_i \cdot h_i \qquad (2.12)$$

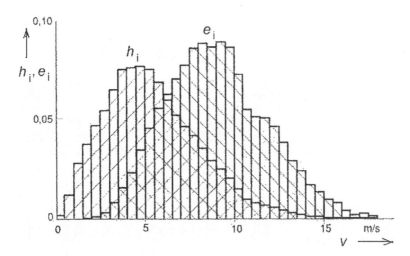

Fig. 2.9 Histograms of wind velocity distribution and normalized energy yield

2.3 Wind Regime and Utilization

Due to the cubic relationship between power and wind velocity the maximum of $e_i(v_i)$ is seen at significantly larger values v_i than the maximum of $h_i(v_i)$. Figure 2.9 gives an example, using a class width of $\Delta v = 0.5\,\mathrm{m/s}$.

2.3.3 Power and Torque Characteristics

The power delivered by a wind turbine is a function of tip speed ratio (see Fig. 2.3) and hence depends on wind velocity and rotational speed. Figure 2.10 shows a normalized representation of power and torque of a rotor with fixed blade position over speed, with wind velocity values as parameter. In the example the rated wind velocity is 12 m/s. The design is such that at approximately $v = 8\,\mathrm{m/s}$ the tip speed ratio is optimal, $\lambda = \lambda_{\mathrm{opt}}$. The power maxima are indicated by the cubic function of v. From the power characteristic it is seen that $P/P_N = 1$ at $n/n_N = 1$. Measures for power limitation at higher wind speeds are not considered in the graphs.

The graphs are analytically derived from a power coefficient curve, see Fig. 2.3. When $c_p(\lambda)$ is given as an empirical function, e.g. in form of a look-up table, then the curves shown in Fig. 2.8 can be calculated by the following algorithm for a specific design.

Let

- λ_A the design tip speed ratio and $c_{pA} = c_p(\lambda_A)$ the optimum power coefficient of the rotor,
- λ_N the tip speed ratio for rated condition, with $c_{p,N} = c_p(\lambda_N)$,
- v_N the wind speed for rated condition.

The ratio λ_N/λ_A can be chosen. In case of a constant speed system for operation at $n/n_N = 1$ the wind speed at which c_p is optimum is calculated by:

$$v_A = \frac{\lambda_N}{\lambda_A} v_N \tag{2.13}$$

To calculate power and torque curves, selected parameter wind speeds v_i are chosen. The following equations are normalized to give P, T and n referred to rated values.

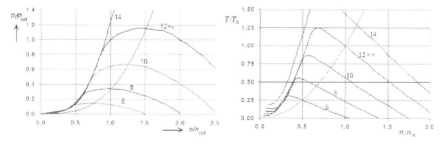

Fig. 2.10 Power and torque characteristics vs. rotational speed ($v_N = 12\,\mathrm{m/s}$)

$$\frac{P}{P_N} = \frac{c_p}{c_{p,N}} \cdot \left(\frac{v_i}{v_N}\right)^3 \quad ; \quad \frac{T}{T_N} = \frac{c_p/c_{p,N}}{\lambda/\lambda_N} \cdot \left(\frac{v_i}{v_N}\right)^2 \quad ; \quad \frac{n}{n_N} = \frac{\lambda}{\lambda_N} \cdot \frac{v_i}{v_N} \quad (2.14)$$

The base quantities for rated condition are as follows:

$$P_N = \frac{\rho}{2} A \, c_{p,N} \, v_N^3 \quad ; \quad T_N = \frac{P_N}{\Omega_N} = \frac{\rho}{2} A \frac{D}{2} c_{T,N} \, v_N^2 \quad ; \quad n_N = \left(\frac{60}{\pi D}\right) \lambda_N v_N$$

where ρ, A are as in (2.2), and c_T as in (2.4); n is in min^{-1}.

The example of Fig. 2.10 is based on the $c_p(\lambda)$ curve of Fig. 2.6, from which $\lambda_A = 6,5$ and $c_{p,A} = 0,52$. The rated wind speed is chosen $v_N = 12\,\mathrm{m/s}$, and the optimum c_p is assigned to $v_A = 8\,\mathrm{m/s}$. According to (2.13) the tip speed ratio at rated condition should be $\lambda_N = \lambda_A(v_A/v_N)$; it was chosen $\lambda_N = 4,5$, with $c_{p,N} = 0,448$.

2.4 Power Characteristics and Energy Yield

2.4.1 Control and Power Limitation

2.4.1.1 Pitching Mechanism

Variation of the blade angle is a means to control the rotor torque and power from the wind side, and at the same time provide power and speed limitation at high wind velocities. Normally the pitch mechanism is powered by a hydraulic or electric drive. Pitch controlled rotors prevail in all larger systems.

In rotors equipped with a pitch mechanism the blade angle is adjusted subject to a relevant controller output. Pitching is also used for power limitation at tip speed ratios above a predesigned value by turning the blades out of the wind. With increasing pitch angle the maximum power and torque coefficients are reduced, and the maxima are shifted to lower λ values. The no-load tip speed ratio is reduced, while the torque coefficient shows increased values at starting. The drag coefficient is significantly reduced. Principal curves of a design, based on the reference Fig. 2.6 are shown in Fig. 2.11.

2.4.1.2 Stall Mechanism

For reasons of safety and to avoid overload, limitation of power above a preset rotational speed is required. This can be accomplished by different methods. A simple way is to turn the rotor out of the wind direction, which is done in the Western Mill. For rotors without pitch angle variation the stall-effect can be utilized, where due to a shift from laminar to turbulent air flow a braking effect is created. Figure 2.12

2.4 Power Characteristics and Energy Yield

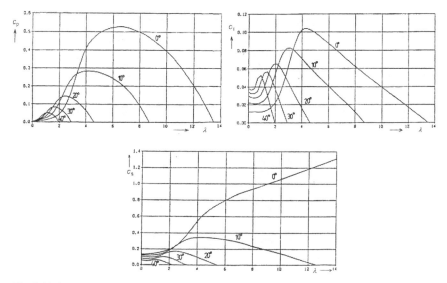

Fig. 2.11 Power, torque and drag coefficients over tip speed ratio with pitch angle as parameter

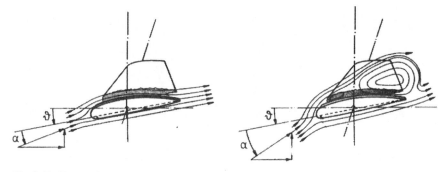

Fig. 2.12 Sketch of a blade with laminar and turbulent air flow

illustrates the effect, where α is the angle between blade plane and oncoming wind direction (as in Fig. 2.4) and ϑ the angle between rotor and blade planes.

A variant of the stall method is the so-called active-stall where the blades are automatically turned into the wind direction when a preset speed is reached. Figure 2.13 illustrates the specific properties. While in stall the air flow break-off occurs with rigid blade position, in active-stall the blades perform a self acting angle variation into the wind to control the air flow breakaway; in pitch eventually the angle variation to decrease the active rotor area is performed without air flow breakaway.

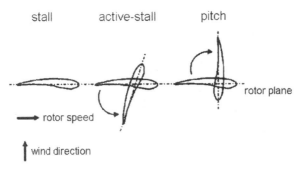

Fig. 2.13 Illustration of stall, active-stall and pitch effects

2.4.1.3 Other Power Limitation Concepts

Small wind turbines in the kW power-range sometimes use other concepts such as passive pitch control, or a passive mechanism tilting the turbine in dependence of wind exerted axial force and decreasing the swept area from circular to elliptic (see 5.2).

Power and speed limitation can also be provided by concepts other than manipulating the wind turbine. This may be an electrical or mechanical braking system.

2.4.2 Wind Classes

Wind energy systems can be assigned to different wind classes. The classes standardized by IEC are commonly used (see Table 2.1). The classes reflect the design dependence on locations with strong or weak wind performance. Characteristic for WES in classes of larger number (lower wind velocities) are larger rotor diameters at same rated power, and often also a larger tower height. Reference values are the average wind speed in hub height and an extremum which statistically happens as 10 min mean value only once in 50 years.

Note that in Germany there is also a classification in wind zones according to the Deutsches Institut für Bautechnik (DIBT).

Table 2.1 IEC type classes

IEC wind class	I	II	III	IV
50 years extremum, m/s	50	42.5	37.5	30
Average wind velocity, m/s	10	8.5	7.5	6

2.4.3 System Power Characteristics

Most important for any wind energy system capability is the power curve. Measured curves of the delivered power over wind speed are, together with the knowledge of average wind velocity and distribution properties (e.g. Rayleigh), indispensable for predicting the annual energy yield. Figure 2.14 shows typical power curves of pitch-controlled and stall-controlled systems. Below a predesigned wind speed, normally the rated wind speed, the power curve is intended to follow a v^3 function using optimum $c_p(\lambda)$. Note that useful power generation starts only at the cut-in wind speed, normally at v between 3 and 4 m/s.

Power limitation at wind speeds above the rated value is effected by either one of the control systems:

- pitch control, where the power is controlled to rated power above a preset threshold wind speed (mostly the rated speed),
- stall control, where a transient phenomenon with power overshoot is observed for wind speeds above rated value.

A number of characteristic wind velocity values are specified with the design for each wind turbine:

- Average velocity v_{av} (in e.g. 10 m or 30 m above ground, or measured in hub height).
- Optimum velocity v_{opt} at λ_{opt} (best point).
- Velocity at maximum energy yield.
- Value of velocity at which the power limitation begins to work. This point is mostly called the rated wind speed.
- Cut-in velocity, at which the turbine starts to supply power.
- Cut-off velocity, at which the turbine is brought to standstill for safety reasons.
- Survival-velocity, in view of an assumed "once-in-a-century storm".

Regulations require a wind energy converter to have two independent braking systems, where the first serves as the main brake and the second as fixing brake. Shut down must be tripped at a maximum wind speed of usually 25 m/s.

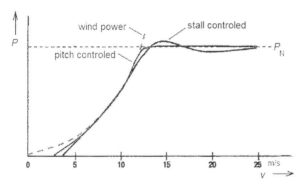

Fig. 2.14 Typical power curves for pitch-controlled and stall-controlled systems

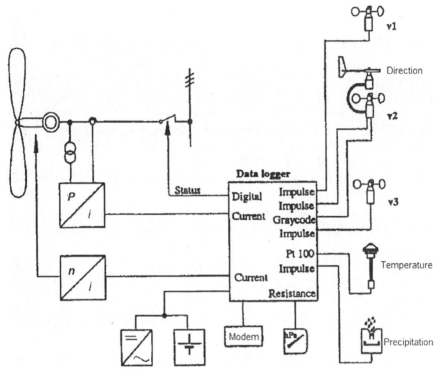

Fig. 2.15 Sketch of a the measuring setup

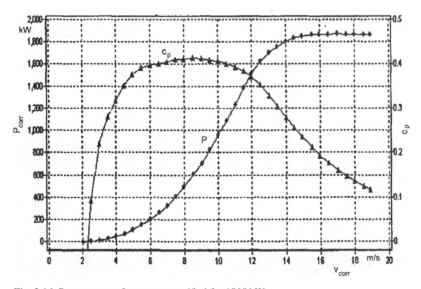

Fig. 2.16 Power curve of a system specified for 1800 kW

2.4 Power Characteristics and Energy Yield

In practice the power curves are determined by test, normally executed in a recognized testing field. Measurements are taken and recorded between cut-in wind speed and cut-off, at least up to 18 m/s. A graph is drawn of the electrical active power P together with the power coefficient c_p versus wind speed v. An example of a turbine rated 1800 kW at 14 m/s is shown in Fig. 2.16, in form of curves of power $P(v)$ and power coefficient $c_p(v)$. The measurement set-up and quantities to be recorded are illustrated in Fig. 2.15 [DEWI].

Figure 2.16 is typical for a design where 14 m/s is the specified rated wind velocity, with a v^3 power dependence below and a limitation above. The power coefficient has its maximum at 8 m/s in a region near the average speed.

A view on specific power values of currently available WES is given in the next two figures, based on data supplied by the manufacturers [BWE07], covering ratings of 850 kW and above. They give an insight into the rotor diameters D and the rotor speeds (upper values in case of speed variable systems) n_m versus rated power, Fig. 2.17. Further, Fig. 2.18 shows the specific power, i.e. the ratio of rated WES power and swept area. According to the trend line the values aggregate between 0,35 and 0,45 kW/m^2, with a tendency to increase with increasing rating. The right part of Fig. 2.18 shows also, for a variety of systems for 2000 kW, the specified specific reference annual energy yield versus hub hight. Here from the trend line values in the regieon 900 ... 1100 (kWh/a)/m^2 can be observed. Note that among the data there are designs for upper as well as for lower average wind velocities.

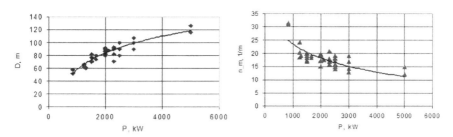

Fig. 2.17 Rotor diameters and rotation speeds of systems of 850 kW and above

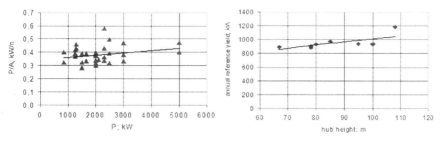

Fig. 2.18 Specific power of systems > 850 kW and reference energy yield of systems 2000 kW

2.4.4 Annual Energy Yield

The annual energy yield depends on the wind speed distribution at the system location in nacelle height and on the machine design parameters. A characteristic parameter is the ratio of the average annual wind speed and the designed rated wind speed. Under certain assumptions (Rayleigh wind speed distribution, design power coefficient $c_p = 0,46$), Fig. 2.19 shows curves of the relative annual full-load energy yield [Gas07]. Considering practical design ratios rated/average wind speed of $1,5 \ldots 2$, it is obvious that characteristic relative full-load equivalent values are not larger than $20 \ldots 35\%$.

A reference energy yield is defined in the German EEG [BRD04]. A reference location is a place with a Rayleigh-distributed wind speed of 5,5 m/s average in 30 m hub height, with a logarithmic elevation profile and 0,1 m roughness length. The

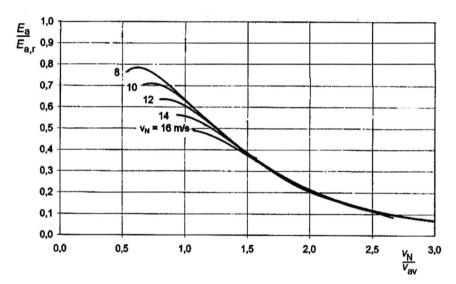

Fig. 2.19 Full-load equivalent annual energy yield over ratio rated/average wind speed

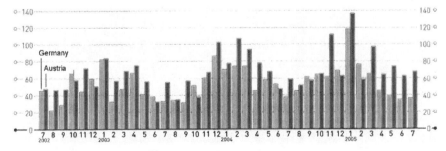

Fig. 2.20 Energy yield kWh/m^2 per month as recorded in Germany and Austria

reference yield is the calculated energy supplied by a specific type of wind system with trhe relevant hub height at the reference location, on the basis of a measured power characteristic over 5 years. Expertise is required that the wind system is capable of at least 60% of the reference yield, in order to receive the established energy price from the utility.

The energy yield per month is of course strongly dependent on the time of the year. For two countries on the Northern Hemisphere, Austria and Germany, reported values of 2002–2005 of onshore wind parks collected in a database are visualized by Fig. 2.20 (Source: Betreiber-Datenbasis and Interessengemeinschaft Windkraft Österreich). It gives an impression of differences not only between months but also between several years. Note that in the meantime the distribution of participating WES has changed to higher unit ratings, together with a better utilization.

Chapter 3
Generators

3.1 General

Electrical machines are usually divided into three groups, namely d.c. machines, asynchronous a.c. machines and synchronous a.c. machines. Of these machine types the d.c. machines are no longer of practical interest as generators because of several drawbacks; they require more maintenance effort, have an unfavourable power to mass ratio and are not suitable for high voltage windings. Of the a.c. machines, both asynchronous and synchronous types are in use. Included are induction machines, denoting asynchronous machines of which only one winding is energized. In most cases the load winding consists of three phases.

The machines discussed in the following belong to the electro-magnetic machines; electrostatic power conversion is outside the scope. The conventional a.c. machine types have an outer stator carrying the main (primary, armature) winding; the rotor is arranged inside the stator, with the air gap separating the inner stator and outer rotor surfaces. The interacting magnetic field crossing the air gap in radial direction couples stator and rotor members. Other constructions, e.g. radial field machines with outer rotor, and axial field machines may be interesting for special designs in the field of wind power conversion.

Since every electrical machine is capable of working as a generator as well as a motor, generators may also run up from electrical supply. The machines can also serve for electrical braking.

A selection of book literature on electrical machines is in [Chap04, Kra02]. The IEC provides a series of international standards applicable to electric machinery in [IEC60034]. Relevant IEEE standards are also mentioned [IEEE112, IEEE115].

In the following asynchronous machines (AM) and synchronous machines (SM) are described regarding their terminal properties and steady-state operation, using equivalent circuit models. Here steady-state means that voltages and currents are sinusoidal functions of a single (fundamental) frequency and the rotational speed is constant. Transient behaviour will be considered in a later chapter.

M. Stiebler, *Wind Energy Systems for Electric Power Generation*. Green Energy and Technology, © Springer-Verlag Berlin Heidelberg 2008

3.2 Asynchronous Machines

3.2.1 Principles of Operation

Asynchronous rotating machines consist of a stator with a preferably three-phase winding, and a rotor carrying either a cage winding or a polyphase coil winding. Normally the stator is the primary member, while the rotor is the secondary member. Induction machine is the term for an asynchronous machine supplied only in the primary part. Cage induction machines prevail in industrial electric drives. Pole pair numbers p of 2, 4, 6 and sometimes 8 are in use, with a preference for $p = 2$ due to advantages in manufacturing and specific cost.

Wound rotor asynchronous machines feature slip-rings and brushes, allowing to feed the rotor winding. This is the case with the doubly-fed asynchronous machines, often used as WES generators.

Figure 3.1 depicts a circuit diagram of a three-phase induction machine with star connected stator winding, the terminals U, V, W supplied from the lines L1, L2, L3 of a three-phase grid. The rotor is also shown with a three-phase winding, the terminals K, L, M connected to slip-rings, the rotor circuit accessible by means of slip-rings and brushes. When the rotor-side is short-circuited the machine is comparable to a cage induction machine. For modeling purposes the cage rotor winding can be transformed into a three-phase winding as shown. Note however, that for cage windings the properties regarding current displacement are different from actual wound rotors, due to the rotor conductor cross section geometry.

Power conversion in asynchronous machines occurs by means of magnetic flux coupling the members via the air-gap. Its fundamental component is rotating with synchronous speed referred to stator frame, With rotor speeds asynchronous to that of the main field, the rotor induced e.m.f. is of slip frequency (causing currents of the same frequency) and a torque is produced.

The machine performance can analytically be described by equations or by equivalent circuits containing lumped parameters of inductances and resistances. Of special interest is the machine behaviour in steady-state, when transients as of switching processes have vanished.

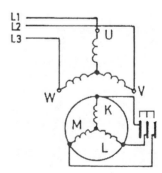

Fig. 3.1 Diagram of three-phase induction machine with wound rotor

3.2.2 Performance Equations and Equivalent Circuits

3.2.2.1 Model Assumptions

Conventional models describe a three-phase machine of symmetrical construction, for operation with the stator winding terminals connected to a symmetrical, sinusoidal supply of constant voltage and frequency. In wound-rotor machines slip power may be recovered in terms of voltages and currents of slip frequency. Analytical equations and respective equivalent circuits serve to determine electric and magnetic quantities for steady-state operation by complex calculus. The use of these models with constant parameters implies:

- constant magnetizing inductance, i.e. neglecting main field saturation effects,
- neglecting eddy-current losses in conducting parts except windings,
- neglecting the influence of current displacement in windings,
- neglecting the influence of temperature variation on resistances.

The conventional T-model in Fig. 3.2 is a two-mesh, per-phase representation of a wound rotor machine, where the rotor winding terminals are accessible via slip-rings and brushes. It shows a specialized form of a transformer model, where the primary (stator) circuit is magnetically coupled with the secondary (rotor) circuit. The primary winding is supplied by voltage $\underline{U}_1$ of frequency ω_1. All secondary side quantities are referred to the primary side by using both a real transformer ratio k representing the windings ratio, and the frequency ratio ω_1/ω_2, where ω_1, ω_2 are stator and rotor frequency, respectively. For cage rotor machines the secondary side mesh is short-circuited, $\underline{U}_2 = 0$.

Figure 3.3 is equivalent to Fig. 3.2 when the machine is supplied by a fixed frequency ω_1, as in grid operation. Inductive parameters are represented by reactances ($X = \omega_1 \cdot L$). Secondary side quantities referred to the primary side are indicated by a prime (') added to the symbol.

In the equivalent circuit representations the parameters are denoted as follows:

R_1; $R'_2 = k^2 R_2$ are primary and secondary side resistances,
L_m is the main field (magnetizing) inductance,
$L_1 = L_m + L_{\sigma 1}$; $L'_2 = L_m + L'_{\sigma 2}$ are primary and secondary side total inductances, the respective leakage inductances being $L_{\sigma 1}$; $L'_{\sigma 2} = k^2 L_{\sigma 2}$,
k is the conventional primary/secondary fundamental winding ratio,
s is the slip as defined in (3.1).

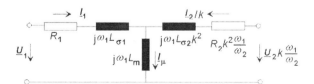

Fig. 3.2 Asynchronous machine T-model circuit

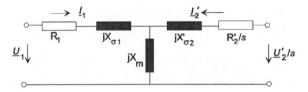

Fig. 3.3 T-model for asynchronous machine connected to the grid

Note the definitions

$$\sigma = 1 - \frac{L_m^2}{L_1 L_2}; \quad k = \frac{N_1 \cdot \xi_1}{N_2 \cdot \xi_2}; \quad I_2' = \frac{I_2}{k}; \quad U_2' = U_2 \cdot k$$

where

N_1, N_2 are the winding turns and ξ_1, ξ_2 the fundamental winding factors of the primary and secondary windings, respectively;

σ is the total leakage coefficient.

Rotational angular speed Ω and slip s are related by the following equation:

$$\Omega = (\omega_1 - \omega_2)/z_p; \quad s = \frac{\omega_2}{\omega_1} = 1 - \frac{\Omega}{\Omega_{syn}} \quad \text{at} \quad \Omega_{syn} = \frac{2\pi f_1}{z_p} = \frac{\omega_1}{z_p} \quad (3.1)$$

where z_p is the number of machine pole pairs, and Ω_{syn} is the synchronous angular speed.

When fed from a source with phase voltage r.m.s. value U_1 of frequency f_1, the voltage equation may be written:

$$\begin{bmatrix} U_1 \\ U_2'/s \end{bmatrix} = \begin{bmatrix} (R_1 + jX_1) & jX_m \\ jX_m & (R_2'/s + jX_2') \end{bmatrix} \begin{bmatrix} I_1 \\ I_2' \end{bmatrix} \quad (3.2)$$

Provisions may be added in the equivalent circuit model to take load-independent (constant) losses approximately into account. To model the iron loss, a resistance R_{Fe} is connected in parallel to the magnetizing reactance, Fig. 3.4a. This is a physically plausible way to model the eddy-current loss due to the main flux oscillations; however the hysteresis loss which follows a different dependency can only roughly be covered by this method. A different but simple method is to model the constant losses (friction, windage and iron losses) by inserting a resistance R_p parallel to the machine terminals, Fig. 3.4b.

Asynchronous machines are capable of self-excitation when, in order to supply the magnetizing current, capacitors are connected parallel to the machine terminals. To model this effect it is necessary to adapt the model by taking the non-linearity due to saturation of the main field inductance into account.

3.2 Asynchronous Machines

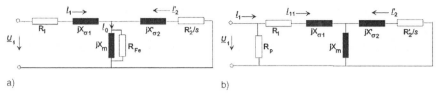

Fig. 3.4 T-model variants containing loss resistors (**a**) conventional iron loss resistor R_{Fe}; (**b**) resistor R_p representing constant losses

3.2.2.2 Operation at Given Stator Voltage

The voltage equation (3.2) can be solved for the currents, when the machine is fed from the grid or a voltage-source inverter of fundamental frequency ω_1. A simplification may be introduced by neglecting stator winding resistance R_1, which is permissible for $R_1/(\omega_1 L_1) \ll 1$. The following equations describe the complex stator and rotor currents as well as the torque. They come in two equivalent representations:

a) with slip s as variable, using short-circuit reactance X_k and breakdown-slip s_k as parameters.

$$\underline{I}_1 = \frac{-j\underline{U}_1}{X_k} \frac{\sigma + js/s_k}{1 + js/s_k}; \quad \frac{\underline{I}'_2}{\underline{I}_1} = -\frac{js/s_k}{\sigma + js/s_k} \frac{X_m}{X'_2}, \quad T = T_k \frac{2s/s_k}{1+(s/s_k)^2}$$

where $\quad X_k = \sigma X_1; \quad s_k = \dfrac{R'_2}{\sigma X'_2}$

b) with the rotor frequency as variable, using rotor time constants τ_{k2} (short-circuit) and τ_{02} (no-load) as parameters.

$$\underline{I}_1 = \frac{\underline{U}_1}{j\omega_1 \sigma L_1} \frac{\sigma + j\omega_2 \tau_{k2}}{1 + j\omega_2 \tau_{k2}}; \quad \frac{\underline{I}'_2}{\underline{I}_1} = -\frac{j\omega_2 \tau_{02}}{1 + j\omega_2 \tau_{02}} \frac{L_m}{L'_2}; \quad T = T_k \frac{2\omega_2 \tau_{k2}}{1 + \omega_2^2 \tau_{k2}^2}$$

where $\quad \tau_{k2} = \dfrac{\sigma L_2}{R_2}; \quad \tau_{02} = \dfrac{L_2}{R_2}$

The breakdown-torque is $\quad T_k = z_p \dfrac{3}{2} \dfrac{U_1^2}{\omega_1} \dfrac{1-\sigma}{X_k} = z_p \dfrac{3}{2} \dfrac{U_1^2}{\omega_1^2} \dfrac{1-\sigma}{\sigma L_1}$ (3.3)

When the stator frequency is considered variable, as by converter supply, a simplified control law requires adjusting the voltage according to:

$$\frac{U_1}{\omega_1} = \text{const.} \qquad (3.4)$$

Under this condition the flux linkage $\Psi_1 = U_1/\omega_1$ is approximately kept constant. Then current and torque depend only on rotor frequency, as shown in Fig. 3.5. The

Fig. 3.5 Principal characteristics of current and torque at constant flux linkage

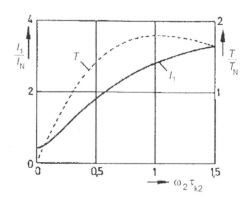

abscissa is scaled in normalized rotor frequency ($\omega_2 \tau_{k2}$). It can be observed that at $\omega_2 = 1/\tau_{k2}$ the developed torque is the break-down value T_k.

3.2.2.3 Grid Operation

When a stator voltage of constant frequency is given, as in grid operation, alternative expressions to calculate the complex currents and torque are preferred. Different from (3.2) and (3.3) reference parameters are now the values of stator-side short-circuit reactance X_k and break-down slip s_k, again neglecting stator resistance.

$$\underline{I}_1 = \frac{U_1}{jX_k} \frac{\sigma + js/s_k}{1 + js/s_k}; \qquad \frac{\underline{I}'_2}{\underline{I}_1} = \frac{j s/s_k}{\sigma + j s/s_k} \frac{X_m}{X'_2} \qquad (3.5)$$

where $X_k = \sigma X_1$: $\quad s_k = R'_2/(\sigma X'_2)$

$$T = T_k \frac{2s/s_k}{1 + (s/s_k)^2} \quad \text{where} \quad T_k = z_p \frac{3}{2} \frac{U_1^2}{\omega_1} \frac{1-\sigma}{X_k}$$

The current formula amended for taking stator resistance into account is given by:

$$\underline{I}_1 = \frac{U_1}{jX_k} \frac{\sigma + js/s_k}{(1 + \rho_1 \cdot s/s_k) + j(s/s_k - \sigma \rho_1)} \quad \text{where} \quad \rho_1 = R_1/(\sigma X_1) \qquad (3.6)$$

Here ρ_1 is a primary damping coefficient

3.2.2.4 Complex Locus and Vector Representation

In the equations of current (3.5, 3.6) the slip s can be considered as independent parameter. With the definition:

$$s = \frac{\omega_2}{\omega_1} = 1 - \frac{z_p \cdot \Omega}{\omega_1} \qquad (3.7)$$

3.2 Asynchronous Machines

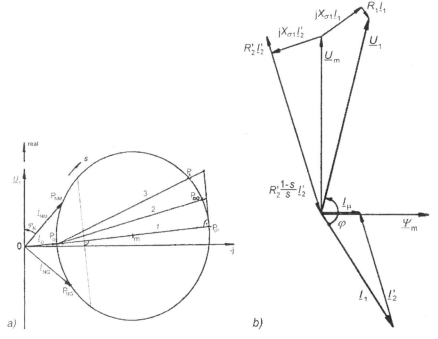

Fig. 3.6 Performance in steady-state (**a**) current locus (Ossanna's circle diagramme); (**b**) vector diagramme (Generator operation)

The current locus curve for steady-state operation can be drawn in the complex plane. For the machine modelled as in Fig. 3.2, and with secondary winding short-circuited, this is a circle. Figure 3.6a shows the so-called Ossanna diagramme. The circle may be defined by three points:

- P_0 for no-load operation, $s = 0$, where current $\underline{I}_0$ is mainly inductive with only a small active component,
- P_k for short-circuit, $s = 1$; where current I_k is typically 5 ... 8 times the rated value,
- P_∞ for unlimited positive or negative slip, rendering a purely inductive secondary side impedance.

The circle diameter is $(P_0 P_\Phi)$ and its centre point m, which is located slightly above the horizontal axis. Note that in the complex plane points of motoring operation are in the upper half-plane, while generator operation is in the lower half-plane. All points of operation are in the right half-plane indicating reactive current consumption (inductive currents), characteristic for an induction machine.

In the figure P_{NM} denotes the rated current in motor operation, showing a phase displacement φ_N with respect to the voltage vector $\underline{U}_1$. In generator operation the rated current is $\underline{I}_{NG}$, as indicated by point P_{NG}.

In Figure 3.6a applying to a machine with short-circuited rotor winding, where current-displacement in conductors is neglected, the current locus is a circle. The figure may also serve for a graphical method to determine power and torque values. Consider a machine operating at complex primary current $\underline{I}$ corresponding to point P on the circle. Of the indicated straight lines, 1 is a diameter defined by the no-load point P_0 and the circle centre m. 2 is the line for zero air-gap power and torque, respectively. 3 is the line of zero mechanical power, when friction and windage losses are neglected. The torque is then proportional to the length of a line (P, line 2), drawn perpendicular to the diameter 1. Similarly, the mechanical power is found proportional to the length of a line (P, line 3), drawn perpendicular to the diameter 1. It is understood that scale factors of power and torque have to be derived from the scale factor for current, using the constant values of U_1 and ω_1, respectively, and taking the per-phase nature of the equivalent circuit model into account.

Figure 3.6b is a vector diagram, indicating approximately rated generator operation. It reflects the circuit equations according to Kirchhoff's law. Note that the slip is negative, with a magnitude of a few percent or below. The main field voltage $\underline{U}_m$, appearing at the reactance X_m in the equivalent circuit, is leading the magnetizing current $\underline{I}_\mu$ by a right angle. In view of the physical properties, the rotor-side resistive parameter may be separated into two components,

$$\frac{R_2'}{s} = R_2' + R_2'\frac{1-s}{s},$$

i.e. by the actual resistance and a slip-dependent component called the useful resistance which may be positive for $1 > s > 0$, negative for $-1 < s < 0$. The diagram takes care of this distinction, showing the ohmic voltage drop and the torque-building voltage component which is in the generator case in opposite phase with the current.

Figure 3.7a shows the performance characteristics of a typical induction motor for industry drives application (90 kW, frame size 280, four poles) when supplied by

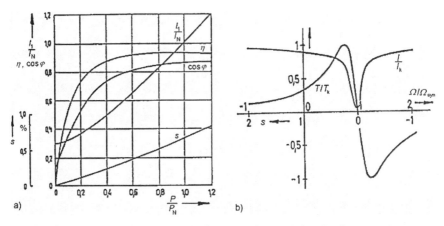

Fig. 3.7 Operation at rated voltage (**a**) Performance characteristics; (**b**) Current and torque vs. speed and slip

3.2 Asynchronous Machines

the grid. Given are normalized current, efficiency and power factor over normalized output power. Note also the slip curve, showing a rated value below 1%. Figure 3.7b is a representation of current and torque over speed in the interval between inverse synchronous speed ($s = 2$), with indication of short-circuit at standstill ($s = 1$), and up to double synchronous speed ($s = -1$). In the torque curve the effect of current displacement is neglected, hence the graph applies to a wound-rotor machine with secondary side short-circuited.

The power behaviour of an induction machine can be described taking the different loss types into account. Below this is done using the air-gap power P_δ which is transmitted electromagnetically between stator 1 and rotor 2.

$$\begin{aligned} P_\delta &= P_{el} - P_{cu1} - P_{Fe} - P_{add} \\ P_{mech} &= P_\delta(1-s) - P_{fw} \\ P_{el} &= P_\delta + P_{cu1} + P_{Fe} + P_{fw} \end{aligned} \quad : \quad (3.8)$$

where

$P_{cu1} = 3 R_1 I_1^2$ stator copper losses (in machines with three-phase windings),
$P_{cu2} = s \cdot P_\delta$ rotor I^2R-losses,
P_{Fe}, P_{add}, P_{fw} iron loss, additional load loss, friction and windage loss, respectively.

The power and loss situation may be visualized by the Sankey-diagram, which in Fig. 3.8 is shown for a motor (a) and a generator operation (b). P_{cu1}, P_{cu2} and P_{add} are the load losses, whereas P_{Fe} and P_{fw} belong to the constant (load-independent) losses. Note that equations 3.11 hold independent of the type of operation as long as the same notation system; here the consumer (motor) system is used, i.e. values P_{el}, P_{mech} are positive in motor and negative in generator operation. The loss values are always positive.

The torque T can be represented by using the air-gap torque T_δ which is related to the air-gap power. To obtain the shaft-torque it has to be diminished by the friction torque T_{fw} which represents the friction and windage losses.

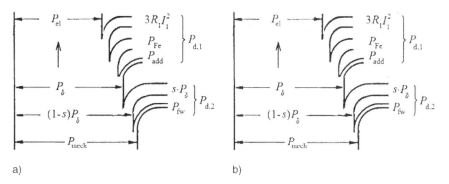

Fig. 3.8 Sankey diagrams of induction machines (a) motor operation; (b) generator operation

$$T_\delta = \frac{P_\delta}{\Omega_{syn}} = \frac{P_{mech} + P_{fw}}{\Omega_{syn}(1-s)}$$
$$T = T_\delta - T_{fw} = T_\delta - \frac{P_{fw}}{\Omega_{syn}(1-s)}$$
(3.9)

Eventually the efficiency can be calculated:

$$\eta_M = \frac{P_{mech}}{P_{el}}; \quad \eta_G = \frac{P_{el}}{P_{mech}} \quad \text{for motor and generator operation, respectively.}$$

3.2.2.5 Per Unit Representation

Often quantities are given in per unit; this is useful for comparing performance or checking parameters of electrical machines and drives. This has been done in Fig. 3.6 for current and torque. The concept is to refer a set of independent state variables to reference values, yielding both variables and parameters as ratios in dimensionless form, expressed as fraction or in %. The method is widely used in the field of power generation and distribution. Preferably rated values are chosen as the reference values.

Use of per unit representation was made for a wind energy converter in the algorithms of 2.3.3.

For application with the induction machine, expressing voltage and currents as r.m.s. quantities and assuming a three-phase star-connected winding, the reference values are U_N, I_N, together with the rated frequency f_N or the related angular frequency $\omega_N = 2\pi f_N$. Knowing the number z_p of pole pairs, the synchronous angular speed Ω_N is also known from (3.1). Then the derived reference values of impedance Z_N, apparent power S_N torque T_N are as follows:

$$Z_N = \frac{U_N}{I_N}; \quad S_N = 3\,U_N I_N; \quad T_N = \frac{S_N}{\Omega_N}$$
(3.10)

Also a reference flux linkage Ψ_N and a reference inductance L_N can be defined:

$$\Psi_N = \frac{U_N}{z_p\,\Omega_N}; \quad L_N = \frac{\Psi_N}{I_N}$$
(3.11)

Usually quantities in per unit are denoted by lower-case letters, e.g.

$$r_1 = \frac{R_1}{Z_N} = \frac{3R_1 I_N^2}{S_N}; \quad x_k = \frac{X_k}{Z_N} = \frac{X_k I_N}{U_N}$$

This means that r_1 is the ohmic loss in the winding resistance referred to rated apparent power; and x_k is the voltage drop on the short-circuit reactance at rated current referred to rated voltage.

3.2.2.6 Operation at Given Stator Current

Consider the case when the machine is fed by impressed currents or from a current-source inverter, terminal voltage and torque are calculated, using the same equivalent circuit model, as follows:

$$U_1 = j\,\omega_1 L_1 \cdot I_1\, \frac{1+j\omega_2\,\sigma\tau_{02}}{1+j\omega_2\tau_{02}}$$

$$T = T_k\, \frac{2\omega_2\tau_{02}}{1+\omega_2^2\tau_{02}^2} \quad \text{where} \quad T_k = z_p \cdot 3\, I_1^2 \cdot L_1(1-\sigma) \quad (3.12)$$

Normally operation with impressed current is from an inverter, preferably under current control maintaining constant rotor or stator flux. Care must be taken for limiting the flux in this type of operation. This requires to take the effect of main field saturation into account, for which the value of magnetizing current I_m is indicative:

$$\frac{I_m}{I_1} = 1 - \frac{L_m}{L_2'}\, \frac{j\omega_2\tau_{02}}{1+j\omega_2\tau_{02}} \quad (3.13)$$

Due to main field saturation the flux is a non-linear function of the magnetizing current, $\Psi_m(I_m)$; according to $\Psi_m = L_m I_m$, the saturation can be alternatively modelled by the non-linear inductance $L_m(I_m)$. Hence in the equations (3.12, 3.13) the parameters are saturation dependant.

Figure 3.9 shows typical curves of the main field flux and inductance vs. magnetizing current in normalized representation.

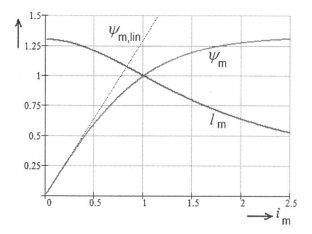

Fig. 3.9 Flux and inductance under main field saturation

3.2.3 Reactive Power Compensation

Asynchronous machines develop their flux by drawing reactive (magnetizing) power via the terminals. In case of sinusoidal voltages and currents the power factor is:

$$cos\varphi = \frac{P_1}{S_1} = \frac{P_1}{\sqrt{P_1^2 + Q_1^2}} \qquad (3.14)$$

Here Q_1 is the fundamental reactive power and S_1 the apparent power. In grid supply $cos\varphi$ can be improved (increased) by a compensation device. In case of stand-alone operation all the required reactive power must be supplied independently.

In grid-supplied a.c. drives it is well known to compensate part of the inductive reactive power by using capacitors parallel to the stator winding, to achieve a power factor of $0,9 \ldots 0,95$. In a three-phase star-connected circuit, using a per-phase capacitance C, the reactive power is:

$$Q = 3 \cdot U_1^2 \cdot \omega_1 C \qquad (3.15)$$

A stepwise adjustable reactive power can be obtained by using capacitor banks. To obtain a continually adjustable compensation, different concepts are applicable. Figure 3.10a indicates the basic circuits.

1. phase-controlled a.c. inductive load and capacitor.
   The voltage across the inductance is adjustable by phase-control, yielding adjustable reactive (inductive) power. Together with the set-value reactive (capacitive) power of the capacitors the resultant reactive power is adjustable and may be designed to cover both signs of reactive power (a1).

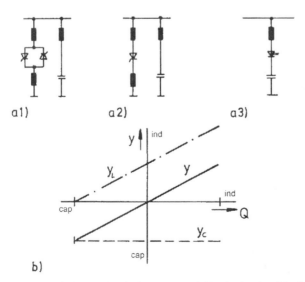

Fig. 3.10 Compensation device concepts (**a**) basic circuits; (**b**) admittance characteristics

3.2 Asynchronous Machines

2. phase-controlled rectifier with inductive energy storage and capacitor
   On the d.c. side of the rectifier an inductor is the storage element. Under phase-control the rectifier consumes a reactive load which is adjustable, and amounts together with the set-value capacitor to a resultant reactive power as under a1 (a2).
3. self-controlled inverter with capacitive energy storage
   An active front-end inverter with an inductor on the a.c. side and a capacitive storage device on the d.c. side is capable of supplying and absorbing reactive power (a3).

Figure 3.10b illustrates in principle characteristic admittances over reactive power in a four-quadrant graph. Note that admittances are defined as follows:

$$Y = \frac{1}{Z}; \quad Y_L = \frac{1}{j\omega_1 L}; \quad Y_C = j\omega_1 C \tag{3.16}$$

In the figure inductive admittance values are positive, the capacitive ones negative.

3.2.4 Self-Excited Operation

While in grid operation the reactive power required for magnetizing the machine is supplied by the utility, different source must be available for island operation.

The classical option is to use capacitors connected to the machine terminals which allow self-excitation. When driven and running on no-load, the machine is acting as an inductance drawing lagging current while the capacitive load supplies a leading current:

$$|I_{10}| = \frac{U_1}{\omega_1 L_1}; \quad |I_c| = U_1 \cdot \omega_1 C \tag{3.17}$$

Self-excitation of the machine is possible due to its non-linear magnetization characteristic. This is illustrated by Fig. 3.11 in normalized representation. At rated

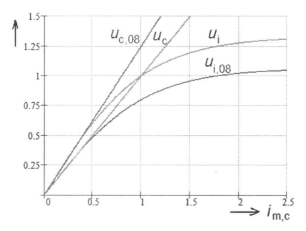

Fig. 3.11 Induction machine capacitive self-excitation

speed the induced voltage $u_i(i_m)$ is analogue to the flux curve in Fig. 3.7; dependence on the speed is shown in the curve $u_{i,0.8}(i_m)$ for 80% rated speed, as an example. The linear capacitor voltage function $u_c(i_c)$, for rated speed and a selected capacitance value, has an intersection with the machine voltage curve at a stable working point which is unity in the figure. The machine becomes a self-excited induction generator (SEIG). Note that the course of attaining this voltage requires an initial remanent flux which is normally present in the machine. When, with unchanged capacitance, the speed is reduced to 80% rated value, $u_{c,0.8}(i_c)$ applies and there is no working point and no self-excitation any longer. The same is observed when at rated speed the capacitance value is reduced to 80%. Consequently for each speed there is a minimum capacitance required for self-excitation.

Figure 3.12a shows a setup where 1 is a driving motor, 2 is the cage induction machine, 3 is the capacitor bank, 4 is a switch and 5 an adjustable ohmic-inductive load. Part b of the same figure is a simplified per-phase equivalent circuit where the induction machine is represented by the so-called gamma-equivalent model and the stator winding resistance neglected. Given the magnetization characteristic and the speed Ω, see (3.1), voltage and currents can be calculated by a suitable iterative method.

Calculated and measured characteristics of a an example setup with a 7,5 kW, 4 pole, 60 Hz induction machine under resistive load are shown in Fig. 3.13. Note that without control measures the terminal voltage strongly decreases with increasing load, and becomes unstable at a specific load dependent on speed and capacitance. The breakdown reflects the point of maximum available capacitive current as difference between capacitor and machine reactive currents at a given voltage (see Fig. 3.11).

The strong dependance of voltage on capacitance would suggest to use variable compensation. However a stepwise switching of capacitor banks to regulate the voltage would not give acceptable results, neither static nor dynamic. Several proposals have been made to realize adjustable voltage control for SEIG. A simple solution is to add in Fig. 3.12a a compensation device as shown in Fig. 3.10a1, using a phase-controlled inductive load parallel to a capacitor. While this is a possible way to achieve a satisfactory voltage controlled stand-alone supply, modern solutions using self-controled inverters are far more preferable, see [Chat06].

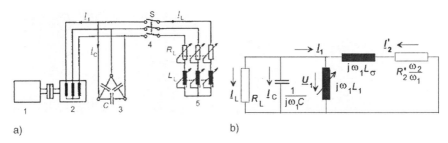

Fig. 3.12 Self-excited induction generator with passive load (**a**) Circuit diagram; (**b**) Equivalent circuit with ohmic load only

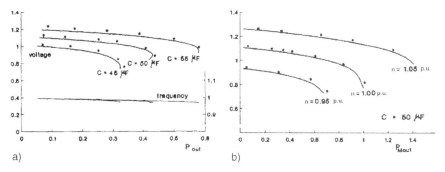

Fig. 3.13 Load characteristics of a laboratory setup with SEIG and ohmic load (**a**) curves at $n = $ const, capacitance C as parameter; (**b**) curves at C = const. speed n as parameter

3.3 Synchronous Machines

3.3.1 Principles of Operation

Synchronous machines as considered here feature a stator carrying a three-phase winding, called the armature winding. The rotor, traditionally called the inductor, supplies the magnetic flux. This is done either by an excitation winding or by permanent magnets. D.c. excitation current is normally transmitted to the rotor by slip-rings and brushes. Concepts without slip-rings, where the excitation current is supplied by a coupled polyphase exciter and rotating semiconductors are known, but are common only in turbo-generators for power stations.

Figure 3.14 depicts a circuit diagram of a three-phase synchronous machine with star connected stator winding, the terminals U, V, W supplied from the lines L1, L2, L3 of a three-phase grid. The rotor carries the field winding, the terminals F1, F2 connected to be fed by d.c. current via slip-rings and brushes from a separate d.c. source with lines L+, L−. The inductor flux may also be supplied by permanent magnets which in the model replace the excitation winding. Note however that this implies fixed rotor flux, different from the case with excitation winding where the flux is adjustable by supplying field current from a suitable source.

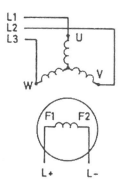

Fig. 3.14 Diagram of three-phase synchronous machine with separate excitation

3.3.2 Performance Equations and Equivalent Circuits

3.3.2.1 Model Assumptions

Synchronous generator inductors are built either with salient poles or as turbo-type rotors. The inductor field generated by either excitation winding or permanent magnets assigns a specified direction to the rotor called the pole axis. Apart from round rotor machines the saliency gives rise to a variable, periodic air gap reluctance long the circumference.

Usually the two-axis theory, also called Park' theory, serves to describe the synchronous machine behaviour. It features an orthogonal coordinate system with the axes d (pole axis, direct axis) and q (intermediate axis, quadrature axis).

The steady-state performance can be discussed using equivalent circuits of lumped parameters. For the case of a turbo-type machine with constant air-gap, In this case equal reluctance in q- and d-axis provides reactance symmetry, $X_q = X_d$. Figure 3.15 shows two equivalent circuits suitable to describe stady state at synchronous speed, one with ideal emf and series impedance, the other with ideal current source and parallel impedance. In the figure $\underline{U}_p$ is the emf induced by the rotor flux (inductor voltage), while $\underline{I}_f$ is the equivalent field current referred to armature side. Note that $X_d = X_{md} + X_{\sigma 1}$.

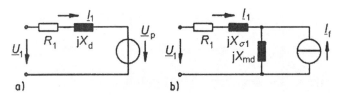

Fig. 3.15 Equivalent circuits of the turbo-type synchronous machine (**a**) with voltage source $\underline{U}_p$; (**b**) with current source $\underline{I}_f$

3.3.2.2 Operation at Given Stator Voltage

In synchronous machine analysis a quantity of special importance is the load

When the machine is running at synchronous speed, $\Omega = \Omega_{syn} = \omega_1/z_p$ with impressed terminal voltage $\underline{U}_1$, the voltage and torque equations of a three-phase turbo-type machine are given by:

$$\underline{U}_1 = R_1\underline{I}_1 + jX_d\underline{I}_1 + \underline{U}_p \quad : \quad \underline{U}_p = jX_{hd}\underline{I}_f = |\underline{U}_p| \cdot e^{j\vartheta}$$
$$T = -\frac{3\,|\underline{U}_1| \cdot |\underline{U}_p|}{\Omega_{syn} \cdot X_d} \sin\vartheta \tag{3.18}$$

where $X_d = X_{md} + X_{\sigma 1}$ is the direct-axis synchronous reaktance, $\underline{I}_f$ denotes the excitation current referred to stator side, for a magnetically homogenous machine, $X_q = X_d$. Load angle angle ϑ is defined as the electrical angle between terminal

3.3 Synchronous Machines

voltage $\underline{U}_1$ and the inductor voltage $\underline{U}_p$. The load angle is positive for leading inductor voltage, as in generator operation.

Siusoidal voltages, currents and flux are represented by vectors in a complex plane, either in Park's d,q-coordinates or in w,b-coordinates indicating active and reactive components as referred to terminal voltage $\underline{U}_1$. The relation between the coordinates is:

$$\underline{I}^{dq} = I_d + jI_q = j \cdot e^{-j\vartheta} \cdot \underline{I}^{wb} \longleftrightarrow \underline{I}^{wb} = I_w + jI_b = -j \cdot e^{j\vartheta} \cdot \underline{I}^{dq} \qquad (3.19)$$

3.3.2.3 Complex Locus and Vector Representation

Figure 3.16a shows complex currents with relative inductor voltage as parameter, in the wb-plane. For the turbo-type machine the loci are concentric circles with center M. Considering the quadrants in the figure, current vectors in the upper half-plane (quadrants I, II) indicate motoring, while the lower half-plane (quadrants III, IV) indicates generator operation. Also currents in the right half-plane (quadrants I, IV) indicate reactive power consumption ("under excitation"), whereas the left half-plane (quadrants II, III) indicates reactive power supply into the grid ("over excitation"). Note that for a given active load the reactive power adjusts itself depending on the inductor voltage. Reactive power can be set in machines with excitation windings. Shown in the figure is a motor operation, over excited.

A vector diagram describing a generator performance with reactive power delivered is shown in Fig. 3.16b. Note the positive load angle of approximately 30°.

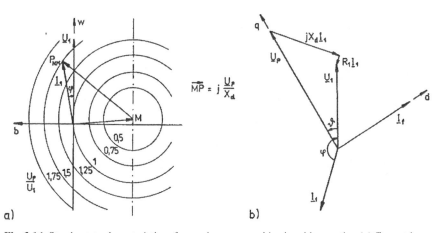

Fig. 3.16 Steady-state characteristics of a synchronous machine in grid-operation (**a**) Current locus diagram; (**b**) vector diagram

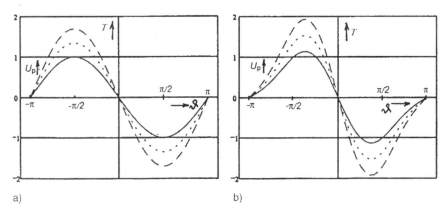

Fig. 3.17 Torque characteristic in grid operation. (a) Turbo-type machine; (b) Salient polemachine, $X_q < X_d$

Figure 3.17a shows the torque vs. load angle. Theoretical limits of stable operation are right angles, $|\vartheta| = \pi/2$, where generating and motoring breakdown torques appear, respectively.

The simple model of Fig. 3.15 is no longer sufficient to describe salient pole machines, where due to magnetic inhomogenity the synchronous reactances in direct and quadrature axis are of different magnitude, $X_q \neq X_d$. The current loci then form Pascal conics instead of circles. A variable-reluctance torque component, proportional to the square of the voltage and the sine of double load angle, is generated in addition to the synchronous component. Figure 3.17b shows an example for a machine where $X_q < X_d$; now the break down torque appearing at $|\vartheta| \leq \pi/2$.

3.3.2.4 Operation at Given Passive Load

Different from grid operation with given terminal voltage and frequency, in stand-alone operation the voltage is load-dependant. Assuming a passive resistive/inductive load as indicated in Fig. 3.18a, and using frequency ω_1 which reflects the speed as independent variable, armature current and terminal voltage are expressed by:

$$\underline{I}_1 = \frac{U_p}{R_{tot}} \frac{1}{1 + j\omega_1 \tau}; \qquad \underline{U}_1 = \underline{U}_p \frac{R_L}{R_{tot}} \frac{1 + j\omega_1 \tau_L}{1 + j\omega_1 \tau} \qquad (3.20)$$

where

$$R_{tot} = R_1 + R_L \,;\; L_{tot} = L_d + L_L \,;\quad \tau_L = \frac{L_L}{R_L} \,;\quad \tau = \frac{L_{tot}}{R_{tot}} = \frac{L_d}{R_L} \frac{1 + L_L/L_d}{1 + R_1/R_L}$$

The inductor voltage (r.m.s. value) is proportional to the impressed inductor flux linkage Ψ_p (amplitude value) and angular speed $\Omega = \omega_1/z_p$:

3.3 Synchronous Machines

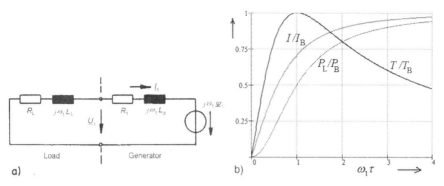

Fig. 3.18 Operation with passive R, L load (**a**) Equivalent circuit for turbo-type machine; (**b**) load curves vs. normalized frequency

$$\underline{U}_p = j\omega_1 \underline{\Psi}_p / \sqrt{2} \qquad (3.21)$$

In the case of a constant load impedance the current magnitude, the torque and the output power are functions of the frequency:

$$|\underline{I}_1| = I_B \frac{\omega_1 \tau}{\sqrt{1 + (\omega_1 \tau)^2}} \quad \text{where} \quad I_B = \frac{\Psi_p}{\sqrt{2} L_d} \frac{L_d}{L_{tot}}$$

$$T = -T_B \frac{2\omega_1 \tau}{1 + \omega_1^2 \tau^2} \quad \text{where} \quad T_B = \frac{3}{2} z_p \frac{\Psi_p^2}{L_d} \frac{L_d}{L_{tot}} \qquad (3.22)$$

$$P_L = P_B \frac{\omega_1^2 \tau^2}{1 + \omega_1^2 \tau^2} \quad \text{where} \quad P_B = \frac{3}{2} \frac{\Psi_p^2}{L_d^2} \cdot R_L \frac{L_d^2}{L_{tot}^2}$$

Normalized curves of current, torque and output power, with reference values I_B, T_B and P_B defined in (3.2) are shown in Fig. 3.18b.

For ohmic load, and assuming $R_1 \approx 0$, output power curves vs. current with speed as parameter are calculated by $P_L = 3U_p I_1 \sqrt{1 - [L_d I_1 / U_p]^2}$ and shown in Fig. 3.19.

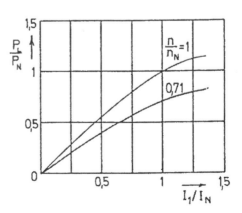

Fig. 3.19 Output power characteristic for ohmic load

3.3.2.5 Permanent Magnet Excitation

Permanent magnet (PM) excitation is used for synchronous machines of increasing rating during the last years and has now been applied for units in the 5 MW class. The absence of excitation winding losses helps to reach high efficiency values. Of the different magnet material technologies (ferrite, AlNiCo, rare earth) today materials of high specific magnetic energy, preferably neodymium-iron-boron, have become standard because air-gap flux densities no less than in conventional machines can be realized, albeit at relatively high cost.

Arrangement of the magnets on the rotor can be on the cylindrical rotor surface (surface-mounted) or in slots (inset magnets) or within the rotor iron (buried magnets), the latter method allowing flux-concentration designs. With surface-mounted magnets the air-gap reluctance is similar to that of a turbo-type machine. With inset and buried magnets there is a variable reluctance torque component; generally $X_d < X_q$.

It is understood that with PM excitation the inductor voltage is not adjustable; of active and reactive power only one is adjustable in grid operation.

3.3.3 Unconventional Machine Types

3.3.3.1 Direct Driven Generators

In systems with direct driven generator there is no gear box in the drive train. This concept allows in principle to reduce maintenance cost and improve efficiency. Closer inspection is needed to evaluate savings in cost and mass.

Systems with synchronous generators for grid supply are always of the speed-variable type, the machine frequency being decoupled from the utility frequency. On the other hand for reasonable exploitation of the machine its rated frequency should not be too much below 50 or 60 Hz. Hence the generator must feature a large number of poles, especially of large ratings. This leads to decreasing power/mass ratios, because according to a basic relation the rated torque (and not rated power) defines the active volume of an electrical machine.

Due to manufacturing considerations, only synchronous machines have found practical application in systems for direct drive while induction generators are confined to systems with gear boxes.

3.3.3.2 Unconventional Designs

Several unconventional designs of synchronous machines have been taken into consideration for wind energy systems. One of the main reasons is to realize constructions of large number of poles in a given volume with reasonable manufacturing effort. Some of them are different from the standard alternating pole concept.

3.3 Synchronous Machines

Axial Field Machines

Axial filed synchronous machines with permanent magnet excitation have been recommended for high torque to volume ratios compared with radial field types. They come as double systems with inner stator or inner rotor. Windings can be ring or trapecoidal coil windings

Figure 3.20 shows principal arrangements. Indicated are in (a) and (b) the rotor-stator-rotor (RSR) and in (c) the stator-rotor-stator (SRS) types. The windings are realized in (a) as ring wound (toroidal) and in (b) and (c) in form of coils: trapecoidal one- or two-layer or concentrated windings. Figure 3.21 indicates, for the RSR-type, typical field lines. In the symmetrical version (a) same magnet polarities face each other, and the stator yoke provides the flux linkage to toroid windings. In (b) is depicted the version suitable for coil windings. The different types have their pros and cons.

Stators can be slotted, with the winding conductors placed in the slots, or slotless with air-gap windings. The latter solution does inherently not produce any cogging torques, which is of advantage in view of application in wind energy systems for starting at low wind speeds. At least for low ratings this may be a preferable solution; see an example of a generator integrated in a small wind systems in Fig. 3.22 [Okl95]. An extension to a machine consisting of a number of modules as shown in Fig. 3.23 has been proposed for large ratings of low-speed synchronous machines, such as ships propeller drives or wind generators [Cari99]. In spite of the claimed advantages design and mechanical problems have hitherto prevented introduction into practical use.

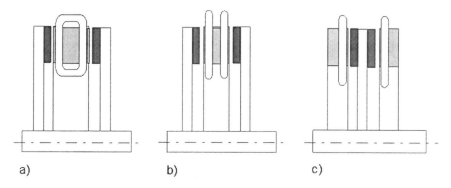

Fig. 3.20 Principal types of axial field machines

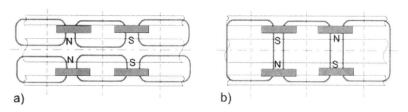

Fig. 3.21 Flux path sketch in axial field machines

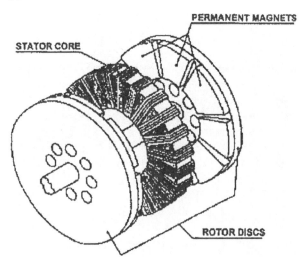

Fig. 3.22 Small axial generator with air-gap ring winding

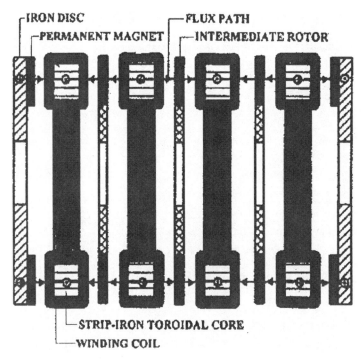

Fig. 3.23 Sketch of modular axial field machine

3.3 Synchronous Machines

Transversal Flux Machines

The principle of transversal flux machines offers advantages producing high force densities and allows realization of large pole numbers [Weh86]. The name illustrates the characteristic flux path which, in a rotating machine, is partly led in axial direction, thus decoupling the space for the armature winding cross-section and the pole pitch. With permanent magnet excitation, transversal flux machines can be designed to force densities up to $100\,kN/m^2$ at linear current densities of $150\,kA/m$, the achievable values increasing at decreasing pole pitch. The concept is considered of advantage for direct drive applications.

On the other hand the 3D construction is mechanically complicated. Small airgaps are required due to the variable reluctance torque component, and the construction tends to give rise to large leakage fluxes; consequently low power factor values are characteristic. Hence up to now their application is scarce due to mechanical reasons and manufacturing cost.

Principal geometric arrangements are depicted in Fig. 3.24. Part (a) illustrates a single-sided version; the permanent magnets are arranged surface-mounted on a soft iron rotor member. The stator ring winding representing one phase is surrounded by U-shaped soft iron elements, placed two pole pitches apart. Flux-leading elements are provided for a closed magnetic path [Hen97]. This version is less challenging in manufacture than the original double-sided concept shown in part (b) of the figure, using an intermediate rotor positioned between two stator elements. Here the permanent magnets are in concentrator arrangement, with alternating polarity and separated by soft iron pieces.

Transversal flux machines were taken into consideration as motors for traction applications; as generators they were proposed for wind energy systems of lower rating already years ago. Recently a concept for a 5 MW direct-driven wind generator was described [Svech06].

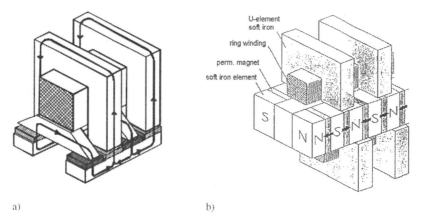

Fig. 3.24 Principle of transversal flux machines (**a**) single-sided component of polyphase machine; (**b**) Double-sided with intermediate rotor

52 3 Generators

Other Concepts

Other types of machines can be considered for application in wind energy systems. All of them are of the synchronous type, where the induced frequency at given rotational speed is considerably larger than in conventional alternating-pole machines. For ratings above some kilowatts however, these unconventional concepts have only appeared as special products or prototypes. The following types are notable.

Modular magnet machines

In these machines the stator is composed of single modules carrying a coil winding. The rotor carries permanent magnets, surface mounted or in flux concentration arrangement. For vehicle propulsion the machine features preferably an outside rotor.

Variable reluctance machines

The principle design is known from stepping motors and finds application as switched reluctance motors. Evidently this machine can also be used as generator.

3.4 Generator Comparison

From the performance of the machine types discussed in the previous chapters the following conclusions can be drawn:

- The asynchronous machine, especially in the form of cage induction machine, is a robust and low cost generator. In the conventional solution directly coupled to the mains, the required reactive power is drawn from the grid. This constant speed technology may be improved by arranging for a second speed in the pole changing concept (Danish concept), preferably in the ratio 3:2.

  When using the wound rotor asynchronous machine the slip power can be recovered. This was done in the static Kramer system. The modern solution is the so-called doubly-fed asynchronous machine which allows, by means of a converter, to extract or feed power in the rotor circuit. Operation with variable speeds in a ratio of typically 2:1 requires a converter designed for approximately 1/3 of the rated power.

  To use the induction machine as a stand-alone generating system, a controlled supply of the magnetizing reactive power is necessary. This can be done by a self-controled (active front-end) machine-side inverter.

  Principal mechanical limitations prohibit very high pole numbers, say above 8. Hence asynchronous generators in wind systems are normally driven via a gear box.

- In the synchronous machine the required magnet flux is provided by permanent magnets or by excitation current fed into a field winding. In the latter case reactive power and terminal voltage, respectively, are adjustable independent from active load. The "grid-building" property makes the synchronous machine especially suitable as generator for island networks.

For variable speed solutions it is necessary to use a converter designed fort the complete rated power to decouple the frequencies.

Synchronous machines may be designed with large number of poles, to operate directly driven in wind systems without gear boxes. Pole pitch values can farther be reduced by applying special concepts.

Chapter 4
Electrical Equipment

4.1 General

This chapter on electrical equipment deals with the power devices used in wind turbine systems, apart from the generators. Conventional devices are only mentioned, while the techniques most important for modern systems is power electronics and converters. Also of exceptional interest for the grid integration of wind energy systems is the topic of energy storage.

A selection of book literature on power electronic devices is in [Moh95; Heu96]; the IEC provides a series of international standards in [IEC60146]. A survey on current and future storage systems can be found in the Wikipedia [enwi].

4.2 Conventional Electrical Equipment

Different conventional electrical equipment is utilized in wind energy systems for switching on and off, for setting active load and reactive compensation. Here are only noted:

- Transformers and inductors,
- Rheostats and load resistors,
- Passive compensation equipment (capacitors),
- Passive filters.

4.3 Power Electronic Converters

4.3.1 General

In several system concepts the generated active power must be adapted in voltage and frequency to the output or consumer side. This is especially the case in variable speed systems. The devices serving for this purpose are power electronic inverters.

M. Stiebler, *Wind Energy Systems for Electric Power Generation*. Green Energy and Technology, © Springer-Verlag Berlin Heidelberg 2008

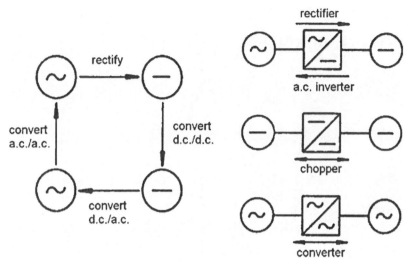

Fig. 4.1 Power electronics applications

Different concepts require different variants of power electronic circuits. Figure 4.1 indicates the tasks fulfilled by power electronic converters [Heu96].

Power electronic devices contain switching elements in form of semiconductors. The semiconductor elements are either not controllable (diodes), or controllable by switching on (thyristors) or by switching on and off (bipolar transistors, MOS-FETs, IGBTs, GTO-thyristors). During operation currents are commutated from one inverter leg to another. Depending on the source of the e.m.f. required for the commutation process self-commutated and external commutated circuitry is distinguished in inverter technology. External sources are grid, machine or a load as source.

Depending on the task the following kinds of inverter are distinguished (see Fig. 4.1):

- *A.c./d.c. inverters (rectifiers)*
  They transform a.c. current of a given voltage, frequency and number of phases into d.c. current.

  Uncontrolled devices contain diodes, normally in bridge arrangement. Most used are the two-pulse (Graetz) bridge for single-phase input, and the six-pulse bridge for three-phase input. Controlled a.c./d.c. inverters are used in external commutated as well as in self-commutated schemes.
- *D.c./a.c. inverters*
  They transform d.c. current into a.c. current of a certain voltage, frequency and number of phases.

  These devices are either external or self-controlled. When the a.c. side is a grid, these inverters can act also as a.c./d.c. inverters, allowing power exchange in both directions.

4.3 Power Electronic Converters

- *A.c./a.c. inverters*

  They transform a.c. current of a given voltage, frequency an number of phases into a.c. current of another voltage, frequency and number of phases.

  Inverters with intermediate circuit combine an uncontrolled or controlled rectifier and an external or self-controlled d.c./a.c. inverter. Both inverters are coupled by the intermediate circuit, either with impressed voltage (voltage-source-inverter, VSI), or with impressed current (current-source inverter, CSI).

  Cyclo converters are built without intermediate circuit; the power is transformed in the same inverter

  A newer concept is the matrix-inverter where fast switches couple input and output side, using no energy storage elements.

- *D.c./d.c. inverters (choppers)*

  They transform d.c current of a given voltage and polarity to d.c. currennt of another voltage and polarity. Inverters using an energy storage element and a pulse-control scheme are usually called choppers.

  Depending on the ratio of output-/input-voltage, we have step-up (boost) inverters for ratios > 1 and step-down (buck) inverters for ratios < 1. Specific inverter circuits are capable of both ways of operation (buck-boost).

Note that in this book resonant and other special inverters which are normally not used in wind energy application are not discussed.

4.3.2 External-Commutated Inverters

4.3.2.1 Thyristor Bridge Inverter

Figure 4.2 shows a fully-controlled rectifier circuit containing a three-phase inverter transformer, a six-pulse bridge (B6) circuit of thyristors and a choke in the d.c. output. Due to the gate-turn-on capability without gate turn-off of the thyristors the commutation voltage must be supplied by the grid. This is the classical a.c./d.c.

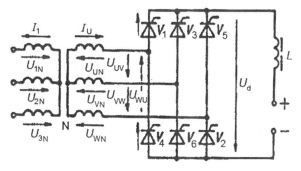

Fig. 4.2 Six-pulse bridge (B6) thyristor rectifier circuit

inverter arrangement known from controlled d.c. drives. The average d.c voltage is adjustable by using phase control, dependent on trigger delay angle α. By firing the thyristor at angle α, commutation is initiated, and the relevant valve takes over load current, until at the next commutation the thyristor is extinguished when the current decreases below the holding current. The circuit as shown is capable of working both as a rectifier supplying d.c. loads from the three-phase grid ($0 < \alpha < \pi/2t$) and as an a.c inverter supplying power from the d.c. to the a.c. side ($\pi/2 < \alpha < \alpha_{max}$), where α_{max} is the maximum permissible trigger delay angle with respect to the stability limit, in practice $\alpha_{max} = 150° \ldots 160°$.

The average d.c. voltage for fully controlled bridge circuit, with $U_2 = U_{UV}$ the secondary line-to-line voltage, is:

$$U_{di\alpha} = U_{di0} \cos \alpha \quad \text{where} \quad U_{di0} = \frac{3\sqrt{2}}{\pi} \cdot U_2 = 1,35 \cdot U_2 \qquad (4.1)$$

Note that the case of a diode bridge rectifier is included with $\alpha = 0$.

Assuming ideal conditions, i.e. neglecting commutation delay and a very large d.c. side inductance, the current waveform will be composed of 120° blocks, see Fig. 4.3. The transformer secondary side current reflects the 120° blocks. The same current blocks appear on the primary side in the case of YY-connection (as shown in the figure), whereas in the case of ΔY-connection the terminal current shows a different step-function.

With I_d the (flat) d.c. current according to Fig. 4.3, the following related quantities are:

– average valve current $\qquad I_v = \frac{1}{3} \cdot I_d = 0,33 \cdot I_d$

– r.m.s. line current $\qquad I_2 = \sqrt{\frac{2}{3}} \cdot I_d = 0,816 \cdot I_d \qquad (4.2)$

– fundamental line current, r.m.s. $\qquad I_{2.1} = \frac{\sqrt{6}}{\pi} \cdot I_d = 0,78 \cdot I_d$

Obviously the currents contain harmonics, of which in the B6 circuit the prominent order numbers are 6, 12 on d.c. side and 5,7,11,13 on the a.c. side.

Figure 4.3 is a simplified picture assuming sinusoidal a.c. voltage, flat d.c. current (no harmonics) and immediate commutation. More realistic is the picture in Fig. 4.4, where commutation delay expressed by an overlap (commutation) angle u is taken into account [IEC60146]. Figure parts (a) and (b) is related to rectifier and a.c. inverter operation, respectively. For the sake of simpler visual representation the waveforms belong to a three-pulse M3 circuit, and not to a B6 circuit as in the former figure.

Presuming an inductance L_C per phase in the commutation circuit, a decrease of the average voltage can be described by using a fictitious series resistance R_x on the d.c. side. Equation (4.1) is to be replaced by:

4.3 Power Electronic Converters

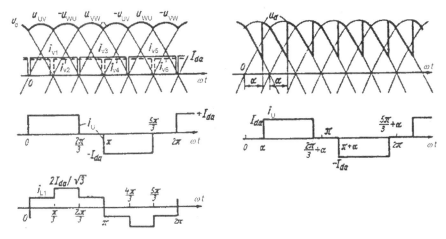

Fig. 4.3 Simplified waveforms of B6 circuit under load (a) Rectifier operation, uncontrolled ($\alpha = 0$) or diode; (b) Rectifier operation, controlled ($0 < \alpha < \pi/2$), example

$$U_{d\alpha} = U_{di0} \cos\alpha - R_x I_d, \quad R_x = \frac{3\omega L_c}{\pi} \quad (4.3)$$

An alternative representation using the commutation angle u is:

$$U_{d\alpha} = U_{di0} \frac{1}{2}[\cos\alpha + \cos(\alpha + u)] \quad (4.4)$$

It is mentioned that the above equations hold for continuous current conduction.

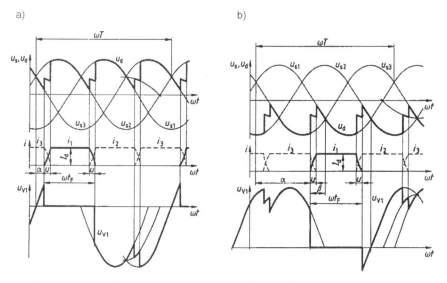

Fig. 4.4 Waveforms of M3 circuit under load, (a) rectifier, $\alpha = 22°$, (b) a.c inverter, $\alpha = 142°$

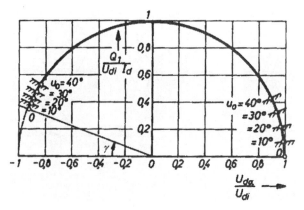

Fig. 4.5 Circle diagram showing the relation between d.c voltage and control-reactive power (u_0 values indicate initial commutation overlap angles)

In the field of wind energy conversion systems the inverter with B6 thyristor bridge is applicable to the classical super-synchronous cascade, identical to the static Kramer system. It must be noted that in variation of the firing angle α not only adjusts the average d.c. voltage, but also a reactive power is drawn from the grid which assumes its maximum (equal to the active power at $\alpha = 0$) when $\alpha = \pi/2$. The reactive power Q is inductive (underexcited). Its fundamental value due to phase control is:

$$Q_{1,c} = U_{di}I_d \sin\alpha \qquad (4.5)$$

When taking the reactive power due to commutation also into account, we get:

$$Q_1 = U_{di}I_d \sqrt{1 - (\cos\alpha - R_x I_d / U_{di})^2} \qquad (4.6)$$

Figure 4.5 shows the relevant circle diagram [Heu96].

4.3.2.2 Reactive Power Current-Source Inverter

While reactive power consumption in circuits with thyristor controlled B6 circuits is mostly considered a drawback because of the poor power factor on the a.c. side when adjusting the d.c. voltage to lower values, a variant of the current source inverter may find application as an adjustable reactive power generator. In the circuit of Fig. 4.6 the d.c. side contains an inductor as storage element. Firing angles of nearly 90° allow to adjust reactive power. Since the commutation voltage must be delivered by the grid side, the device can only generate inductive reactive power.

The circuit is a controllable static var compensator (SVC), of which use is made in FACTS technology for shunt-connected controllers. With regard to wind energy systems the device can serve to control the voltage of a self excited induction generator (SEIG), together with a capacitor bank.

4.3 Power Electronic Converters

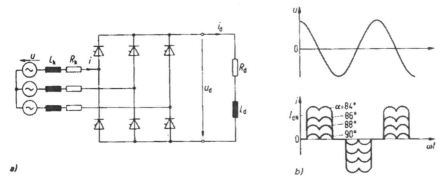

Fig. 4.6 Reactive power inverter with inductive storage element (**a**) circuit; (**b**) voltage and currents (example)

4.3.3 Self-Commutated Inverters

4.3.3.1 Three-Phase Full Wave Bridge Inverter

In self-commutated inverters the commutation voltage must be generated within the power electronic device. Generally this requires semiconductor elements with gate turn-on and turn-off capability, such as MOSFETs, bipolar transistors, IGBTs or GTOs. Figure 4.7 shows a voltage source inverter in three-phase B6 arrangement, characterized by diodes anti-parallel to the switching elements in the branches.

When controlled in full-wave, six-step operation (or as block inverter), the voltages appear as shown in Fig. 4.8, where d.c. voltage U_d is assumed constant. Figure part a) are the voltages at the a.c. output with respect to point 0. Part b) shows the line-to-line voltages, part c) is one of the phase voltages with respect to neutral point N of the transformer winding in Y connection. Eventually figure part d) depicts the voltage between d.c. reference 0 and a.c. neutral N.

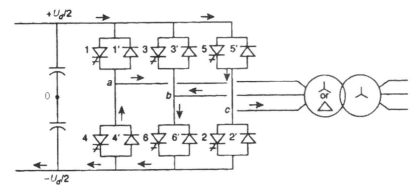

Fig. 4.7 Full-wave bridge voltage-source inverter

Fig. 4.8 Voltage waveforms in six-step operation

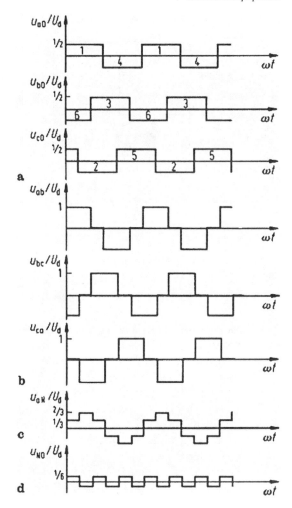

The fundamental r.m.s. value of the three-phase, line-to-line output voltage is:

$$U_{ab} = \frac{\sqrt{6}}{\pi} U_d = 0,78 \cdot U_d \qquad (4.7)$$

Note that in this configuration the output frequency is adjustable by controlling U_d, but the output voltage amplitude is not. The currents results from the interaction of the output voltage with the a.c. side system which may be an induction machine.

4.3.3.2 Voltage-Source PWM Inverter

Figure 4.9 shows a three-phase, two-level converter, with semiconductor elements as in Fig. 4.7, and an ohmic-inductive load at the a.c. output. Output frequency and fundamental voltage can be controlled in switch-mode operation.

4.3 Power Electronic Converters

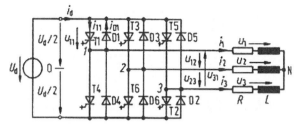

Fig. 4.9 PWM bridge (B6) inverter circuit

Many pulse pattern generation schemes are known, fulfilling different optimization approaches. One of the most popular ones is the pulse width modulation (PWM). A synchronous modulation using sinusoidal reference voltages u_{ref} and a triangular carrier voltage u_{tri} is illustrated in Fig. 4.10. The frequency ratio of the triangular and reference voltage which should be an uneven integer is $m_f = 9$ in the example. Switch pairs connect an output branch either to the positive or negative d.c terminals, depending on whether $u_{ref} > u_{tri}$ or $u_{ref} < u_{tri}$. This is shown in the Figure for u_{10} and u_{20}, resulting in the output line-to-line voltage u_{12}.

The modulation ratio is defined using reference and triangular voltage amplitudes:

$$m = \frac{\hat{u}_{ref}}{\hat{u}_{tri}} \qquad (4.8)$$

In the region $m \leq 1$ the fundamental output line-to-line voltage is a linear function of m:

$$U_{12} = \frac{\sqrt{3}}{2\sqrt{2}} m \cdot U_d = 0{,}612 \cdot m \cdot U_d \quad \text{for} \quad m \leq 1 \qquad (4.9)$$

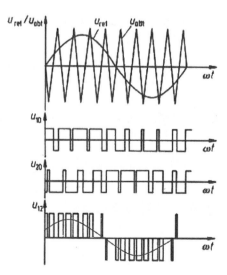

Fig. 4.10 Voltage waveforms in PWM operation with sinusoidal modulation

The region $m > 1$ is called overmodulation, and the output voltage is further increasing in nonlinear way. The maximum is the case of square-wave modulation according to (4.7), which occurs at a value of m dependent on the frequency ratio m_f. The voltage waveform contains harmonics, with foremost order numbers $(m_f \pm 2)$, $(2m_f \pm 1)$ and $(3m_f \pm 2)$, their amplitude values depending on the modulation ratio m.

In a symmetrical three-phase circuit as shown in Fig. 4.9, the phase voltages also form a symmetrical system; Fig. 4.10 indicates the waveform envelope for PWM modulation. The load currents are determined by the load impedances.

4.3.3.3 Active Front-End Inverter

A self-commutated inverter combined with an a.c. side inductor, called active front-end inverter, is often used to couple a three-phase a.c. system with a d.c. circuit such as a voltage-source intermediate circuit. It is capable of conducting active power and of reactive power in both directions, thus allowing control of the power factor. Its operation is discussed using a.c. complex calculation for sinusoidal fundamental quantities.

Figure 4.11 is the model, where compared with Fig. 4.9, instead of a grid there is now an electrical machine on the load side which may be an induction generator. An inductor of impedance $\underline{Z}_k$ couples the machine terminals with the a.c. side terminals of the inverter, $\underline{U}_g$ and $\underline{U}_v$ are fundamental frequency per phase r.m.s. voltages of the generator and the inverter, respectively. A voltage drop at impedance $\underline{Z}_k$ is due to current $\underline{I}$.

Using complex calculus, and fixing vector $\underline{U}_g$ along the real axis, the current can be expressed by:

$$\underline{I} = \frac{U_g}{\underline{Z}_k}\left[\lambda\, e^{j\beta} - 1\right]$$

where $\quad \underline{U}_g = U_g; \lambda = \dfrac{|\underline{U}_v|}{U_g}; \quad \beta = \arg(\underline{U}_v), \underline{Z}_k = R_k + j\omega_1 L_k = Z_k \cdot e^{j\varphi_k}$

In Fig. 4.12 four example cases are illustrated, using the conventional consumer (motor) coordinate system. In the upper row cases active power is supplied to the machine (motor operation), while the lower row cases are for power supplied to

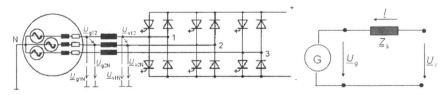

Fig. 4.11 Model of generator, coupling inductor and inverter

4.3 Power Electronic Converters

Fig. 4.12 Phasor diagram of a.c. fundamentals in different cases of operation

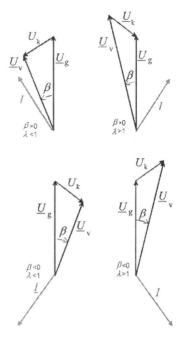

the converter intermediate circuit (generator operation). On the other hand the left column cases denote reactive power supply from the machine side (over excitation, capacitive), while the right column cases indicate reactive power consumption (under excited, inductive).

The complex apparent power of the three-phase system becomes:

$$\underline{S} = 3\,\underline{U}_g \underline{I}^* = S_{ref} \cdot e^{j\varphi_k}\left[\lambda e^{-j\beta} - 1\right] \quad \text{where} \quad S_{ref} = \frac{3U_g^2}{Z_k} \quad (4.10)$$

$\underline{I}^*$ denotes the conjugate complex of $\underline{I}$, and S_{ref} is a reference apparent power. Assuming a lossless inductor impedance, $\underline{Z}_k = j\omega_1 L$, $\varphi_k = \pi/2$, the following approximation is obtained:

$$\underline{S} = \frac{3U_g^2}{\omega_1 L}[\lambda \sin\beta + j(\lambda \cos\beta - 1)] = S_{ref} \cdot j(\lambda\,e^{-j\beta} - 1) \quad (4.11)$$

A further approximation may be given for small values of β; indicating a simple concept to decoupled control of active power (by β) and reactive power (by λ).

$$\underline{S} \approx S_{ref}[\lambda \cdot \beta + j(\lambda - 1)] \quad (4.12)$$

In Fig. 4.13 active and reactive power behaviour is shown depending on ratio λ and angle β for lossless inductor. Examples of operation in each of the four quadrants are reflected by the phasor diagrams of Fig. 4.12. Note that an induction

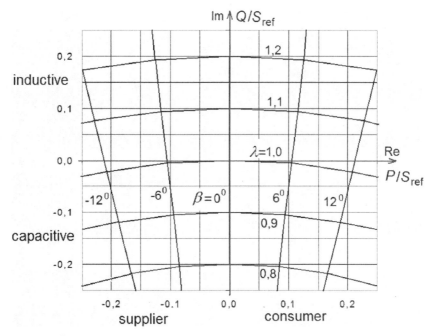

Fig. 4.13 Power locus diagram in consumer (motor) coordinates, for lossless inductor

machine requiring magnetizing reactive power and acting as generator operates in the second quadrant of the figure.

4.3.3.4 Reactive Power Voltage-Source Inverter

Figure 4.14 describes a reactive power inverter with capacitor storage element, shown for square-wave operation [Heu96]. Unlike the current-source inverter described in 4.3.2.2 this self-commutated voltage-source converter is capable of producing capacitive (overexcited) as well as inductive (underexcited) reactive power. The circuit may readily be adapted for PWM operation.

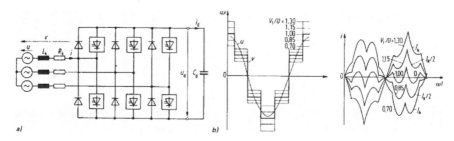

Fig. 4.14 Reactive power inverter with capacitive storage element, in square-wave operation (**a**) circuit; (**b**) voltage and currents (example)

4.3 Power Electronic Converters

The circuit may be considered as a special application of the active front-end inverter, and performance analysis of voltage and current fundamentals derived from 4.3.3.3.

$$U_k = \frac{\omega L_k I_N}{U_N} = 30\% \quad \frac{R_k I_N}{U_N} = 2\% \quad C_d \to \infty \quad t = 50\,\text{Hz}$$

4.3.4 Converters with Intermediate Circuits

Power electronic devices converting a.c. power of one voltage and frequency to another voltage and frequency are known as cyclic converters, matrix converters and converters with intermediate circuit. Considering from the latter those with d.c. intermediate circuit, we distinguish

– voltage-source inverters (VSI, U-converter) with a capacitor storage element;
– current-source inverter (CSI, I-converter) with an inductor storage element.

Figure 4.15 shows principle circuit representations.

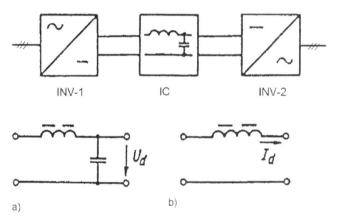

Fig. 4.15 Converter schemes with intermediate circuits. (**a**) voltage-source inverter (VSI); (**b**) current-source inverter (CSI)

4.3.5 D.c./d.c. Choppers

Typical d.c/d.c. converters are shown in Fig. 4.16 [Moh95]. The input voltage U_d is transformed to the adjustable output value U_o, with either a step-down (buck) or step-up (boost) ratio. A suitable semiconductor element such as a transistor, represented in the circuits as a switch S, is periodically switched on and off, controlled by pulse-width modulation (PWM). An inductance L serves as storage element. In normal service, with continuous inductor current, the performance is such that in the buck converter the input current is discontinuous, while in the boost converter

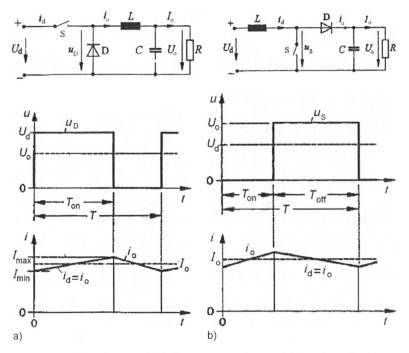

Fig. 4.16 D.c./d.c. chopper with inductor storage element. (a) Step-down (buck) converter; (b) Step-up (boost) converter

this applies to the (output) inductor current, dependent on conduction of switch and diode. The performance of both converters is described under simplified conditions (losses neglected) by Fig. 4.16. It is seen that the inductor is periodically loaded and unloaded, the currents piecewise linear with time. To limit the current variation, a minimum inductance value has to be applied which is inversely proportional to the switching frequency $f = 1/T$. Characteristic values are as follows:

		Buck	Boost
Duty ratio	$d = T_{on}/T$	$d = k$	$d = (k-1)/k$
Voltage ratio	$k = U_o/U_d$	$k = d$	$k = 1/(1-d)$
Current variation	$\Delta I = I_{max} - I_{min}$	$\Delta I = \frac{U_d T}{L} k(k-1)$	$\Delta I = \frac{U_d T}{L}\frac{k-1}{k}$
Required inductance		$L \geq \frac{U_d T}{\Delta I_{max}} k(k-1)$	$L \geq \frac{U_d T}{\Delta I_{max}}\frac{k-1}{k}$

$$\text{(4.13)}$$

Figure 4.16 shows examples with continuous currents. Under different condition the current may be discontinuous. Here the step-up converter is considered to discuss the limits between operation modes [Moh95]. In Fig. 4.17a is shown the limiting case where the decreasing inductor current i_L is touching zero before it

4.3 Power Electronic Converters

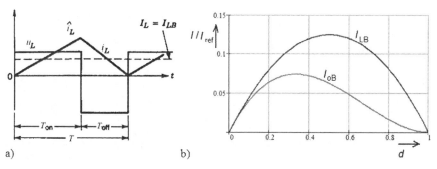

Fig. 4.17 Step-up converter, continuous and discontinuous conduction. (**a**) bondary case of cont./discont. conduction; (**b**) boundary curves; $I_{ref} = U_o T/L$

increases again. The assigned average current I_{LB} is depending on the duty ratio d. Figure 4.17b shows boundary curves of the output current I_{oB} and the inductor current I_{LB}, when the output d.c. voltage U_o is kept constant. The following maximum values apply:

$$I_{LB,\max} = \frac{1}{8}\frac{U_o T}{L} \quad \text{at} \quad d = 0{,}5\,;\quad I_{oB,\max} = \frac{2}{27}\frac{U_o T}{L} \quad \text{at} \quad d = 1/3 \quad (4.14)$$

Note that the step-up converter is not suitable for no-load operation; the output voltage must always be larger than the input voltage by at least 2 V. Control characteristics are given in Fig. 4.18, again for output voltage $U_o = $ const, indicating the limit between continuous and discontinuous conduction.

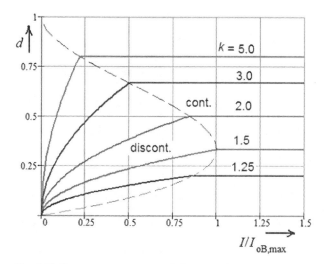

Fig. 4.18 Step-up converter, control characteristics

Regarding wind energy systems, d.c./d.c. choppers, when applied, are mostly boost converters, e.g. in so-called battery-loaders of low rating.

4.3.6 A.c. Power Controllers

A.c. power controllers are phase-controlled electronic devices, which provide adjustable r.m.s voltage of a series-connected load at unchanged frequency. They come in single-phase and three-phase devices, using semiconductor elements in form of thyristors arranged in anti-parallel connection or triacs which conduct both a.c. half-waves in one element. Figure 4.19a shows a typical circuit with ohmic-inductive load, the neutrals being not connected. Dependent on control angle α the following part of the voltage-time-area is applied to the load and current begins to flow, ending at the moment when the current through the semiconductor element reaches zero. Since no commutation phenomena occur, the power controllers are usually not called converters.

The r.m.s voltage applied to the load can be adjusted from full wave at $\alpha = 0$ downwards to zero. In a three-phase controller, Fig. 4.19a, this is the case at $\alpha = 150°$. The current is a function of the load, in the example it depends on its characteristic phase angle $\varphi = \text{atan}(\omega_1 L/R)$. Figure 4.19b shows the current control characteristics for selected cases $\cos \varphi = 1$ and $\cos \varphi = 0$; for other load phase angles the curves start at $\alpha = \varphi$ with maximum current $I_{0,\text{rms}}$. Load voltage as well as current contain harmonics, their amplitudes depending on α. Hence additional losses occur, e.g. when the load is an induction motor.

In wind energy systems such power controllers are used for connecting induction generators to the grid, to avoid rush-currents in so-called soft-start devices.

Phase controlled thyristor power controllers are also used as a.c. switches without mechanical contacts.

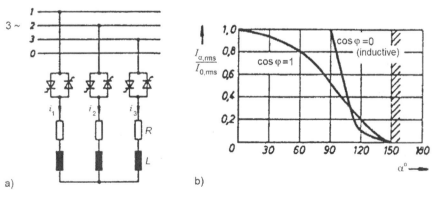

Fig. 4.19 Three-phase power controlers (**a**) Circuit with ohmic-onductive load; (**b**) Control characteristic of r.m.s. current

4.4 Energy Storage Devices

4.4.1 General

Storing electrical energy is important issue in different technical fields. In power generation and distribution pump-storage plants are used to supply energy in periods of peak-load, and to pump water back in periods of weak load. A wide field to use energy storage devices is in vehicles.

Different technologies are in use or in development. In [enwi] the forms are listed: Pumped water, Batteries, Compressed air, Thermal, Flywheel, Superconducting magnetic energy, Hydrogen. Hence storage methods are electrochemical, electrical and mechanical. The devices allow reversible energy conversion in both directions, by charging and discharging, their properties making them suitable for either short-time or long-time storage of energy. Energy storage capacity and power capability, as well as the specific values of energy density and power density, are characteristics for applying and comparing different technologies. In the following a selection of methods is discussed, as far as they are or may be considered in connection with wind energy application.

The natural variations in wind speed and consequent power fluctuations give rise to a wide interest in energy storage techniques. The increasing percentage of power from wind parks fed into the grid requires measures for temporary and regional balance of generation and consumption. Regarding wind energy generation in island grids energy storage devices also play an important role for the grid concept. Eventually, in low power wind systems such as battery loaders a storage device is constitutive. Hence it is justified to consider storage technologies from large energy and power ratings down to low ratings.

Comparison of the capabilities of different storage devices in terms of energy density and power density can roughly be made from the Ragone diagram in Fig. 4.20. Fuel cells will not be considered here.

4.4.2 Electrochemical Energy Storage

4.4.2.1 Lead-Acid Batteries

Lead-acid batteries are the classical accumulators for application both in vehicles and in stationary equipment. The accumulators used in solar and wind conversion systems are modifications of the conventional batteries for automobiles.

One cell of the battery has two electrodes: the negative one of lead, Pb, and the positive one of leadoxide, PbO_2. The electrodes are submerged in sulphuric acid, H_2SO_4. They have the following reversible reaction:

$$Pb + PbO_2 + 2H_2SO_4 \longleftrightarrow 2PbSO_4 + 2H_2O$$

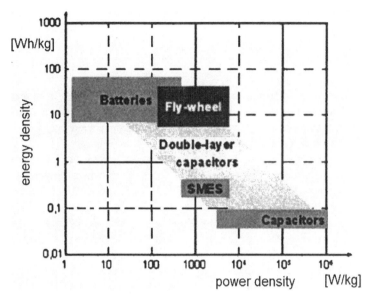

Fig. 4.20 Ragone diagram of energy storage devices

When charged, and at 25 °C the acid density is $1,24\,\mathrm{g/cm^3}$. The cell voltage is roughly 2 V, resulting in 12 V for a battery with 6 cell in series, a number common from batteries for passenger cars. In fully charged state the voltage is somewhat larger, at 2,3 V. Figure 4.21 shows the cell voltage over degree of charge, indicating

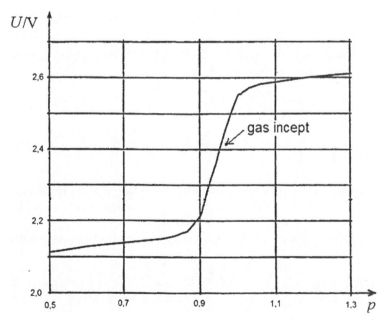

Fig. 4.21 Cell voltage U over degree of charge p

4.4 Energy Storage Devices

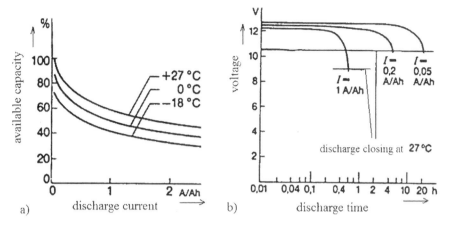

Fig. 4.22 Characteristic of a conventional battery (**a**) Available capacity; (**b**) Battery voltage during discharging

also the inception of gas secretion at 2,4 V. Hence the nominal voltages of commercial accumulators are 14 V, 28 V and 42 V. The nameplate battery capacity is given in Ah.

The available capacity depends mainly on temperature and the drawn discharge current. A common definition is the capacity C_{20}, meaning that the specified capacity is available when the battery is discharged with an output current I_{20} such that the discharge closing voltage is reaches after 20 h. The closing voltage is normally 10,5 V, and below for larger discharge currents. Figure 4.22 illustrates the characteristics of a standard 12 V battery. Overcharging and deep-discharge should be avoided.

Different charging methods are known, suitable for vehicle and stationary batteries, for normal or fast charge. One of the most common methods is using the IU-characteristic; Fig. 4.23. In the first interval the current is kept constant the voltage reaches 2,4 V per cell and gas secretion occurs, continuing with constant voltage until full capacity is reached.

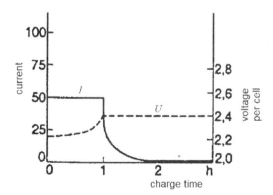

Fig. 4.23 Charging method using I/U control (example)

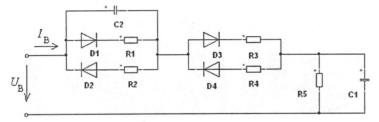

Fig. 4.24 Battery equivalent circuit

Several circuit models were proposed for simulation purposes, different in effort and accuracy. One of them is discussed here shortly [Sal92]. Figure 4.24 shows the equivalent circuit. The model components represent:

C1	battery main capacitance
R5	self-discharge resistance
C2	overvoltage capacitance
R3, R4	internal resistance for charge and discharge, respectively
R1, R2	overvoltage resistance for charge and discharge, respectively

The generally nonlinear parameters are determined from tests. For simulation the properties are modelled by functions, e.g. by $U_{C1}(Q_{C1})$ for C1 or $I_{R3}(U_{R3})$ for R3. For practical use the resistance values are described in sums as R1+R3 and R2+R4. Temperature variation may be taken into account by a temperature compensation algorithm.

4.4.2.2 Other Secondary Batteries

Electro-chemical storage devices apart from the lead-acid accumulator are known for either higher energy densities, longer life cycle or aptitude for quick charge. On the other hand they are more costly. In wind energy systems they have not found much application, but a few of them which are of general interest are mentioned below for comparison. Table 4.1 lists the main properties, with lead-acid cells included for completeness.

NiCd-batteries are widely used in appliances and notebooks. Metals are nickel and cadmium, the electrolyte being caustic potash dilution. They are problematic due to the detrimental effect of cadmium for the environment. Besides, they show the memory-effect, i.e. an irreversible decrease of capacity occurring when the battery is not fully charged in suitable intervals.

NiMH-batteries are less problematic regarding environmental impact. Metals are nickel, titan, vanadium, circonium and chrome, the electrolyte is caustic potash dilution. Compared with NiCd-batteries they have the same cell voltage of 1,2 V, but a larger energy density, and no memory-effect is observed. On the other hand their temperature operating range is more limited.

NaS-batteries feature a higher energy density; however the high operation temperature and the chemically active component natrium were responsible for the fact

4.4 Energy Storage Devices

Table 4.1 Properties of accumulators

	Lead-acid	NiCd	NiMH	NaS
Energy density, Wh/kg	$25\ldots35$	$35\ldots55$	$50\ldots80$	$80\ldots150$
Power density, W/kg	100	200	300	$90\ldots160$
Operating temperature, °C	$-10\ldots60$	$-20\ldots55$	$-20\ldots45$	$290\ldots350$
Cell voltage, V	2	1,2	1,2	2,1
Self-discharge, %/month	$5\ldots15$	$20\ldots30$	$20\ldots50$	0
Useful full cycles	$600\ldots1500$	~2000	~1000	~1000
Energy efficiency, %	$70\ldots85$	$60\ldots75$	$65\ldots85$	$80\ldots95$

that their development was discontinued, and the once intended vehicle application has come out of sight.

4.4.3 Electrical Energy Storage

4.4.3.1 Double-Layer Capacitors

Conventional capacitors are a storage element for electrostatic energy in principle; as devices for storing energy in larger amounts they are not suitable. This limitation has been overcome by the development of electrochemical double-layer capacitors (EDLC). They are known under different trade marks, e.g. UltraCaps, SuperCaps or BoostCaps. EDLCs have two electrodes and a fluid electrolyte. When a voltage below that of electrolyte decomposition is applied, ions collect on the electrodes of opposite polarity in layer thickness of only a few molecules. The very low gaps together with large electrode areas lead to very high values of energy density. The electrostatic energy is:

$$E_q = \frac{1}{2} C U^2$$

Cell capacitances of 5000 F and above are commercially available, with rated voltage/limiting voltage of 2,5/2,8 V; allowing energy densities of up to 4,7 Wh/kg. The useful energy is determined by the difference of the squared voltages charged and discharged, the latter set e.g. at 1,2 V. The series resistance is very low, so that power densities of up to 10 kW/kg can be reached. Balancing the series connected cells requires some equalizer effort, and power electronic devices are indispensable for adaption to the supplying system. On the other hand EDLCs have a long lifetime, allowing 500.000 charging/discharging cycles. Although their energy density is below that of electrochemical accumulators, with their easy handling, good operating temperature range $(-35\ldots65\,°C)$ and high reliability they are becoming a favourite choice for storage applications [Ha196] and especially for vehicles [Schn02]. Regarding wind energy systems they are also considered as short-time storage elements; in [Kin04] a compensation device using an EDLC bank with a buck-boost converter is proposed.

4.4.3.2 Superconducting Storage Devices

Superconducting magnetic energy storage (SMES) systems store magnetic field energy by a d.c. current flowing in a coil which is cryogenically cooled down below its critical temperature by means of a refrigerator. In superconducting state the coil resistance is zero, so that no current decay occurs and theoretically the magnetic energy is stored indefinitely. The magnetic energy is:

$$E_m = \frac{1}{2} L \, I^2$$

The design of coils for SMES has to take into account the inferior strain tolerance of conductor material, thermal contraction due to cooling and the Lorentz forces due to the currents. Structural support is required for the coils which are built either in solenoid or toroid form. The solenoids are less costly to manufacture, and preferred for small SMES. The toroid, on the other hand, creates less external magnetic force, needs less support effort and is therefore the choice for larger coil sizes.

The system consists of the superconducting coil, a cryogenically cooled refrigerator, a power electronic rectifier/inverter to convert a.c. power to the coil arrangement (charging) and to convert d.c. power back to the a.c supply (discharging), and a protection device for emergency discharging. SMES systems are suitable for dynamic operation. High overall efficiency values e.g. 95% are reached.

Since their invention high temperature superconductors (HTSC) are an option beside low temperature superconductors (LTSC). Since for LTSC the typical basic temperature is 4,2 K and the medium liquid Helium, for HTSC the respective value is 77 K and Nitrogen as medium. Note that the critical current density of the wire for HTSC is lower than for LTSC. Evidently the refrigeration cost depends also on the technology used. An indicator is the losses due to heat conduction in the support structure and terminal leads and to radiation, and its ratio to the input power of the refrigerator.

As an example for a SMES system with LTSC wire of NbTi for 2 MJ ACCEL has reported the following properties: Stored energy 2,1 MJ, current 1000 A, coil inductance 4,1 H, flux density 4,5 T, converter intermediate d.c. voltage 800 V, nominal discharge power 200 kW for A > 8 s, maximum power 800 kW.

A historical overview on SMES is in [Luo96]; modeling and simulation is discussed in [Ars99].

4.4.4 Mechanical Energy Storage

4.4.4.1 Water Pump Storage

Pumping water into a higher reservoir in times of low power demand, and utilize the water flowing back to the reservoir on lower elevation through turbines when the demand is high, is a well developed energy storage method. The machines may be a set of motor-generator coupled to a pump and a turbine, for operation in one dense of rotation. The other option is to use a motor-generator coupled to a hydraulic machine acting as pump or turbine, requiring reverse direction of speed. In the latter method the resulting efficiency will be lower with synchronous machines operating at grid frequency. This disadvantage is overcome when by means of a converter the speed is adjusted in pumping operation where the machine is running asynchronously [Hin00]. Conversion losses and evaporation losses from the exposed water are responsible allow in practice for overall energy efficiencies of 70...85%.

The energy density of pump storage systems is relatively low. The potential energy of a water volume V in a reservoir with a water head H (ρ = water density; g = earth surface acceleration) is described by:

$$E_h = \rho g \, H \, V$$

Note that $1 \, m^3$ of water elevated to $100 \, m$ represents a potential energy of $0,277 \, kWh$. Consequently large volumes of water elevation differences are necessary to store significant values of energy. Of course an appropriate geography is most important for any water pump storage project.

A large number of pump storage systems, exploiting an upper reservoir with either a lower reservoir or a flowing stream have been built all over the earth, beginning around 1930. The individual rated power is covering a vast range of $80...1800 \, MW$.

A closer look may be thrown on the Goldisthal plant (2002). The upper reservoir has a volume of $13,5 \cdot 10^6 \, m^3$ representing $8489 \, MWh$. There are 4 machine sets, 2 of them synchronous for $256 \, MW$ each in pump and turbine operation; the other 2 are asynchronous machines for $265 \, MW$ each. The total rated power is specified $1060 \, MW$.

Water pump storage is considered a means for equalizing the long time fluctuations of wind energy systems.

4.4.4.2 Compressed Air Storage Devices

Reversible compression and expansion of air is used in compressed air energy storage (CAES) devices. Air is compressed by a motor driven compressor and stored in an underground reservoir, preferably using caverns created by salt-deposits. When discharging the compressed air is used to burn natural gas in combustors. In modern concepts the heat gained during compression is used for prewarming. The resulting combustion gas is then expanded in a two-stage gas turbine which drives the generator. The usual configuration is a one-shaft assembly with a single electric machine as motor-generator unit. Figure 4.25 shows the concept.

Compressed air storage has been identified as a good combination to supply control energy in grids with a large percentage of installed wind energy systems. At present only a few CAES plants are in practical use. Already commissioned in 1978

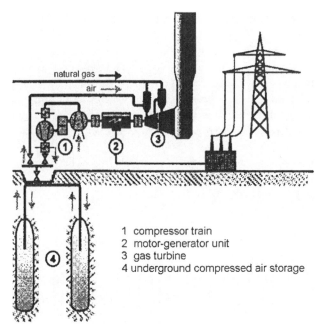

Fig. 4.25 Concept of a CAES plant (Huntorf/Germany)

and still operated by E. On is the plant in Huntorf/Germany whose specification serves here to indicate the system capability, see Table 4.2 [Crot01].

The overall efficiency is reported 54%; full power is available within 10 min.

Table 4.2 Main properties of CAES plant Huntorf

Output	in turbine operation	290 MW for $<=$ 3 hrs
	in compressor operation	60 MW for $<=$ 12 hrs
Cavern pressure	43 bar (minimum)	70 bar (permissible)
Reduction rate	15 bar/h (maximum)	
Total cavern volume	310000 m^3 (2 caverns)	

4.4.4.3 Flywheel Storage Devices

In flywheel energy storage (FES) use is made of the kinetic energy of a rotating device which is:

$$E_k = \frac{1}{2}J\Omega^2$$

where J is moment of inertia referred to the center of rotation, and Ω is the angular velocity. In case of a rotating solid cylinder of mass m and radius R we have

4.4 Energy Storage Devices

$J = m \cdot R^2/2$, and in case of a cylindrical ring of mass m, inner and outer radii R_i and R_a, respectively, we get $J = m \cdot (R_a^2 + R_i^2)/2$. The equation indicate the preference for large rimmed rotors to obtain large values of J, and for high speeds Ω. The rotor may be a steel flywheel, however carbon-fiber composites are of advantage regarding tensile strength. Flywheels have to be specified for a maximum permissible outer surface speed.

In FES systems, the rotor is accelerated to the permissible speed (charging), and decelerated for discharging, using a motor-generator set and a power electronic inverter. No-load losses in bearings and gas friction have to be minimized in flywheel storage devices. Vacuum enclosures and magnetic bearings are used to meet these requirements.

Advanced flywheels are rated for speeds of 20000 min^{-1} and above. High energy densities of 130 Wh/kg and overall energy efficiency of 95% are reached. These properties and long lifetime make them an attractive however costly choice not only in stationary systems such as uninterruptible power supplies (UPS), but also in vehicles. Flywheels have meanwhile also found interest as load levelling devices acting together with wind energy systems.

Chapter 5
Wind Energy Systems

5.1 General

In some applications the rotor operates within power and speed limits on its natural characteristic, e.g. with a centrifugal pump as load. Power limitation in rotors with rigid blade angle may be effected by passive acting measures such as spoilers or braking flaps. For machines of small rating also passive pitching mechanisms, e.g. by means of pre-loaded springs, are known. Speed limiting methods acting on the load side such as feeding energy into a dump resistor are also restricted to small ratings.

Systems for electrical power generation of other than small ratings will generally require active control measures. While pitching allows speed control on the wind side, suitable generator systems provide control possibilities on the electrical side.

Operation management has the task to provide an efficient, failure-free and safe operation. It controls switching on and off of generator and other electrical systems, auxiliary systems such as hydraulic pitching equipment and braking procedures. The task implies state monitoring and surveillance of the protective devices. This requires continuous measuring of weather data, operational and grid data, and processing in the operation management computer.

5.2 Systems Overview

5.2.1 General

In conversion systems to supply electrical energy the wind rotor is coupled, generally via a gear box, with an electrical generator in form of an induction machine or a synchronous machine. In variable speed systems power electronic equipment is used to decouple voltages and frequencies of generation and grid side.

Wind energy systems may be operated according to the following concepts:

- constant speed (as for mains fed synchronous machines), or almost constant speed (as for the shunt characteristic of mains fed induction machines),
- variable speed (as for generators with frequency decoupling by inverters).

M. Stiebler, *Wind Energy Systems for Electric Power Generation*. Green Energy and Technology. © Springer-Verlag Berlin Heidelberg 2008

Note that mains fed synchronous machines mentioned here for completeness are not used in practice. Variable speed systems are found both with induction and synchronous generators.

A conventional concept is to couple an induction generator output directly to the grid, in a system for operation at almost constant speed. A variant is the so-called Danish Concept, where the generator has two winding systems of different pole numbers, e.g. 4 and 6. This allows operation at two specific speeds in the ratio 2:3, where the lower speed serves to utilize the lower wind velocities.

According to the wind rotor power characteristic the optimal operation point shifts to larger speed values as the wind velocity increases. This shows the superiority of variable speed concepts where the rotational speed is adjustable. Variable speed systems for grid supply require adaptive devices between generator and grid; these are generally inverters using power electronics.

In principle the same applies to systems for stand-alone operation. However, in this case frequency and voltage regulation, and namely the reactive power requirements need special attention.

5.2.2 Systems Feeding into the Grid

Systems for feeding into a 50 Hz or 60 Hz network are coupled to a medium voltage or high voltage connecting point. Under normal conditions the frequency may be considered constant, and voltage variations are within specified values, e.g. $\pm 6\%$.

Figure 5.1 shows circuits of typical system concepts. Using an induction generator IG, the electrical machine can be directly coupled forming a system of almost constant rotational speed (a). Variable speed systems use a converter to decouple generator speed from grid frequency, either fully fed (b) or with the converter only for slip energy recovery (c). The latter requires a wound-rotor, slip-ring induction machine. Systems with a synchronous generator always work fully fed with a

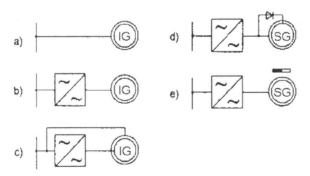

Fig. 5.1 Typical concepts for generating electrical power. – using induction generator; (**a**) direct coupling, (**b**) fully-fed, (**c**) doubly fed, – using synchronous generator, fully fed; (**d**) electrical excitation, (**e**) PM excitation

5.2 Systems Overview

Fig. 5.2 Common concepts of systems feeding into the grid (Legend see text)

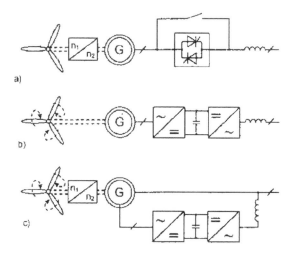

converter; the machine may be electrically excited, via slip-rings or brushless (d), or by permanent magnets (e).

A closer look into the concepts which are realized in the vast majority of wind parks gives. Figure 5.2. Part (a) depicts a conventional system, with an induction generator directly connected, driven by the wind turbine via a gear box, where speed ratios of around 100 are common for ratings of 1500 kW and above. To avoid high rush-in currents after switching, it is usual to have a soft-starting device, consisting of a phase-controlled power electronic circuit. The figure may easily be amended for a generator with two winding systems of different pole pair number, according to the Danish Concept.

Figure part (b) is typical for systems with a synchronous generator, preferably directly driven which implies a design with a large number of pole pairs. Variants are known where a gear box of only moderate speed ratio of around 10 is used (Multibrid), allowing a smaller generator size. The power is fed to the grid via a converter with intermediate d.c. circuit which must be designed for full load (fully fed).

Figure part (c) is the circuit common for a system with doubly fed slip-ring induction generator. In contrast to part figure (b) the converter rating is typically only 35% of full load to allow for a speed range of 1:2. Power electronic adaptive devices are shown as intermediate circuit converters, but other configurations are also possible. Modern equipment uses active front end inverters on the machine side as well as on the grid side, to allow reactive power supply and power factor adjustment.

5.2.3 Systems for Island Supply

Wind energy systems for feeding power into an island grid in absence of a utility may operate in concepts combining different types of generation, such as diesel-generators and photovoltaic units, using also energy storage (buffer) devices. Dependent on the grid configuration, the task of providing a set-value frequency and a controlled nominal voltage has to be fulfilled by one or more components.

Stand-alone wind systems for low rating in the kW-range are known as "battery loaders", where an accumulator is supplied from a directly driven synchronous generator with PM excitation via a diode rectifier and loading controller. Systems with a.c. output may be organized as shown in Fig. 5.2b, where the battery is inserted in the converter's d.c. intermediate circuit. Induction generators may also be used for stand-alone systems, as long as a self-controled (active front-end) machine-side inverter is used to provide the magnetization reactive power.

5.2.4 Wind Pumping Systems with Electrical Power Transmission

A special case of a wind system with electrical output is the one with one load device directly connected, such as a water pumping system, driven by an induction motor. The electric power transmission ("electrical shaft") allows erecting the wind system in a suitable elevated place for supplying a deeper situated pump in a well. The characteristics of a centrifugal pump match favourably with those of the wind rotor, since for both the torque varies in proportion to the speed squared, and simple dedicated control concepts may be used [Mul00].

5.3 Systems for Feeding into the Grid

5.3.1 General

It was already mentioned that, parallel to the increase in installed power in wind parks, the rating per unit has seen considerable growth. Figure 1.3 shows the values for Germany as an example. Statistical values allow a closer look on the distribution of turbine systems within classes of diameter values; Fig. 5.3 applies to the same country for systems installed in 2004 and 2007, respectively [End08]. It is obvious that turbines with pitch control and variable speed systems prevail, especially for larger diameters.

The concepts of constant and variable speed may be discussed in view of the principal characteristics of power and torque over rotational speed. The turbine characteristics in Fig. 5.4 are similar to Fig. 2.8, scaled in normalized values; however wind speed parameter values are in m/s with an assumed rated value of 12 m/s. Variation of the pitch angle is not considered in the Figure. Note that the maximum power values follow a third power curve. The relevant points on the torque curves are found on the falling slopes.

5.3 Systems for Feeding into the Grid

rotor diameter	25 - 45 m		45,1 - 64 m		64,1 - 80 m		> 80 m	
no gearbox	57	6	29	82	414	262	3	110
gearbox	0	0	88	31	407	92	203	300
pitch	57	6	95	113	817	353	134	390
stall	0	0	14	0	0	0	0	0
active-stall	0	0	8	0	4	1	72	20
1 fixed speed	0	0	0	0	0	4	0	0
2 fixed speeds	0	0	27	0	4	1	72	20
variable speed	57	6	90	113	817	349	134	390
number of WTs	57	6	117	113	821	354	206	410

in 2004

in 2007

Fig. 5.3 Shares of type groups installed per year in 2004 and 2007 in Germany

5.3.2 Induction Generators for Direct Grid Coupling

Induction generators directly feeding into the grid are normally cage induction machines. The grid imposes the frequency and supplies the magnetizing reactive power. In normal operation the speed deviates only by slip values of 1% and below from synchronism. Different from variable speed systems, this concept is often called constant speed. Together with a stall controlled wind turbine the system can obviously operate with optimal tip-speed ratio at one single speed only.

Often a second speed is provided in the generator, generally by means of a pole changing winding. Thus the lower occuring wind speeds can be utilized by a second stage of higher pole number at lower rotational speed (Danish Concept). The pole numbers are usually 4 and 6, giving a speed ratio of 3:2. In principle the same effect can be accomplished when a smaller generator coupled used, coupled to the main machine.

The capability of such a wind energy system is characterized by the power curve and the annual energy yield. The latter is defined in dependence of the average wind velocity at hub height. For calculation it is usual to assume a Weibull wind distribution with form factor 2, i.e. a Rayleigh distribution, and constant air density. Figure 5.5 shows an example power curve, based on measurements made in a test field, with uncertainties indicated [IEC61400].

A sketch of such a typical system is depicted in Fig. 5.6. It shows the data acquisition stage where operational and weather data are scanned, and the data processing stage located in a computer which is at the same time the control computer. Indicated is the soft switching device in form of a phase-controlled thyristor inverter, serving to limit the rush-in currents.

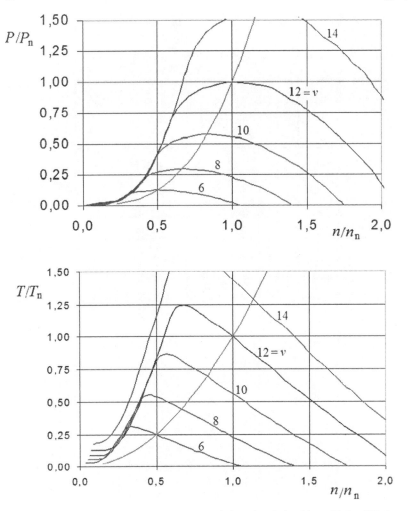

Fig. 5.4 Normalized power and torque characteristics of a wind turbine with fixed blades

Figure 5.7 is an illustration indicating the weather and operational data to be measured by means of suitable sensors, and their use as input quantities to the control routines executed in the computer which has the task of operation management.

5.3.3 Asynchronous Generators in Static Cascades

5.3.3.1 General

Induction machines with wound rotor allow access to the rotor winding via slip rings and brushes. They have been used in grid-supplied drives for heavy starting

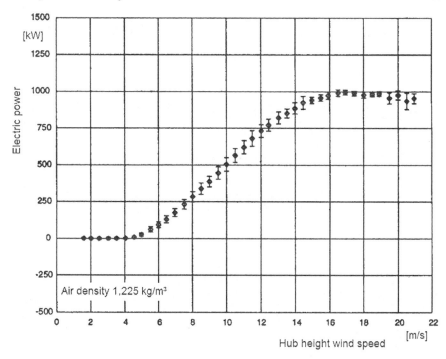

Fig. 5.5 Power curve of a stall turbine system

conditions, where external resistances act as load on the rotor side to increase torque up to breakdown value during starting. Moreover slip-ring rotor machines are suitable for cascade concepts where slip power is recovered or fed into the rotor winding.

An equivalent circuit model of a wound-rotor induction machine is shown in Fig. 5.8, representing a variant of the T-model in Fig. 3.3. It is called the Γ-model for its form of inductance parameter arrangement; another name is L-model. In the sense of linear network theory it is equivalent to the circuit in Figs. 3.2 and 3.3 when using a different transformer ratio:

$$k_\Gamma = \frac{L_1}{L_m} k \tag{5.1}$$

Using this definition, the leakage parameters are concentrated on the secondary side, and equation (3.2) becomes:

$$\begin{bmatrix} \underline{U}_1 \\ \underline{U}'_2/s \end{bmatrix} = \begin{bmatrix} (R_1 + jX_1) & jX_1 \\ jX_1 & (R'_2/s + jX'_2) \end{bmatrix} \begin{bmatrix} \underline{I}_1 \\ \underline{I}'_2 \end{bmatrix} \tag{5.2}$$

where

$$X'_2 = \omega_1 (L_1 + L'_\sigma) = X_1 + X'_\sigma \; ; \quad L'_\sigma = L_1 \cdot \sigma/(1-\sigma) \; ; \quad R'_2 = k_\Gamma^2 \cdot R_2$$

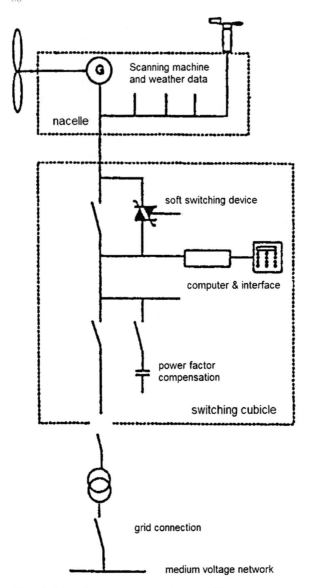

Fig. 5.6 Principle diagram of a constant speed system

Note that in comparison with (3.2) rotor voltage and current referred to stator side are now $\underline{I}'_2 = \underline{I}_2/k_\Gamma$ and $\underline{U}'_2 = \underline{U}_2 \cdot k$.

When investigating different cascade circuits it is useful to define the rotor side voltage in relation to the stator voltage $\underline{U}_1$, using a complex ratio $\underline{\gamma}$, expressed by the real factor k_2 and the phase displacement φ_2:

$$\underline{U}_2' = \underline{U}_1 \cdot \underline{\gamma} = k_2 \cdot \underline{U}_1 \cdot e^{j\varphi_2} \tag{5.3}$$

5.3 Systems for Feeding into the Grid

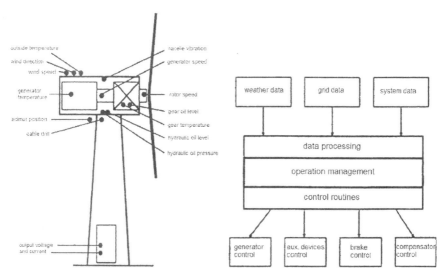

Fig. 5.7 Operation and control of a wind energy system. Sketch of data acquisition (*left*) and block diagram of operation management (*right*)

where $\underline{U}_1$, $\underline{U}'_2$ are the stator voltage and the rotor voltage vector referred to stator side, respectively. Note that in steady state the rotor quantities are of slip frequency $(s \cdot f_1)$.

When neglecting the stator resistance R_1 in Fig. 5.8, which is permissible for most performance calculations not intending to render correctly the machine losses, currents and powers can be determined by a set of complex equations.

Rotor and stator current are calculated:

$$\underline{I}_2' = \frac{1}{X_\sigma} \frac{\underline{U}_2 - \underline{U}_1 s}{s_k + js} = \frac{\underline{U}_1}{X_\sigma} \frac{k_2 \cdot e^{j\varphi_2} - s}{s_k + js}; \text{ where } s_k \text{ is the breakdown slip } s_k = R'_2/X'_\sigma$$

$$\underline{I}_1 = \underline{I}_0 - \underline{I}_2' \quad \text{mit} \quad \underline{I}_0 = \frac{\underline{U}_1}{jX_1} \tag{5.4}$$

Stator and rotor apparent powers are:

$$\underline{S}_1 = 3\underline{U}_1 \cdot \underline{I}_1^* = P_1 + jQ_1 \quad ; \quad \underline{S}_2 = 3\underline{U}_2 \cdot \underline{I}_2^* = P_2 + jQ_2 \tag{5.5}$$

where $\underline{I}^*$ denotes the conjugate complex of $\underline{I}$. Note that a machine with three-phase windings both in stator and rotor is assumed, and $\underline{U}$, $\underline{I}$ are per-phase values.

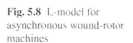

Fig. 5.8 L-model for asynchronous wound-rotor machines

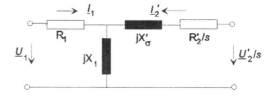

5.3.3.2 Super-Synchronous Kramer System

The doubly fed asynchronous generator system has been mentioned as a standard solution of today. An earlier configuration is the static Kramer system known in the form of the subsynchronous motor cascade in drive technology, but is also applicable as supersynchronous generator cascade. It may serve here to explain the basics of slip power recovery.

The stator winding is directly connected to the grid, while the rotor winding feeds a converter having an intermediate d.c. circuit containing an inductor, see Fig. 5.9. While the machine-side inverter is an uncontrolled diode bridge rectifier, the grid side inverter is a phase controlled thyristor device. The latter is generally connected to the grid via a transformer for voltage level adaption. The converter is designed for slip power conversion.

To investigate the performance of the cascade, the simplified machine model Fig. 5.8 is used, neglecting the stator resistance. It is understood that secondary side quantities are transformed to the primary (stator) side. The grid-side current source inverter is controlled to provide a specific negative d.c. side voltage $U_{d\alpha}$ by setting the phase angle α_W; where $\pi/2 < \alpha_W < (\pi - \beta)$. To secure stability of the inverter performance, the minimal extinction angle β is normally kept about $\pi/6$.

Regarding the machine-side converter, and neglecting any phase displacement between rotor voltage and fundamental rotor current, $\underline{U}_2'$ must be in phase with $\underline{I}_2'$. In this case the imaginary part of the apparent rotor power vanishes, and the following implicit equation applies:

$$s \cdot \cos \varphi_2 + s_k \cdot \sin \varphi_2 = k_2 \tag{5.6}$$

When the rotor voltage is of constant magnitude, i.e. $k_2 = \text{const}$, the same circle applies to the current locus of $\underline{I}_1$, with only a specific slip division characteristic for the given k_2 which is the new no-load slip of the cascade. The current locus circle

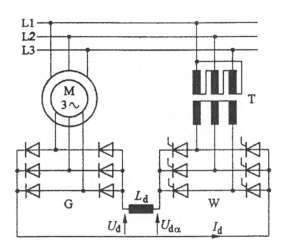

Fig. 5.9 Static Kramer system, basic circuit diagram

5.3 Systems for Feeding into the Grid

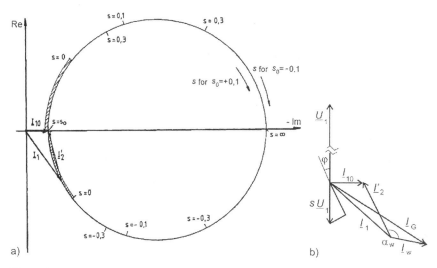

Fig. 5.10 Steady state operation of the wound rotor induction machine (**a**) complex current locus for constant rotor voltage ($s_0 = \pm 0,1$ as example); (**b**) phasor diagram for operation as a generator

in Fig. 5.10a illustrates an example for $k_2 = 0,1 = $ const of a machine having a breakdown slip $s_k = 0,303$. Note that the circle has two slip scales, the exterior for $s_0 = -k_2 = -0,1$, and the interior for $s_0 = k_2 = 0,1$. Quantities s_0 are no-load slip values in cascade operation.

It becomes evident that subsynchronous motor operation and supersynchronous generator application is obtained with the same circuit. In the case of an uncontrolled rotor side rectifier the hatched parts in Fig. 5.10a would require active power supplied to the rotor which cannot be realized when the diode rectifier is used. A principal phasor diagram is shown in Fig. 5.10b for negative slip, $s = -0,35$. As mentioned, phasors $\underline{U}_1$ and $(s \cdot \underline{U}_1)$ are in phase opposition. Note that the magnetizing current $\underline{I}_{10}$, lagging by $\pi/2$, is exaggerated in length in the figure. Rotor current $\underline{I}'_2$ is transformed to the grid side current $\underline{I}'_W$ by action of the inverter, producing a phase angle $\alpha_W \sim 5/6 \cdot \pi$. The resulting generator current is $\underline{I}_G$.

Both supersynchronous and subsynchronous operation can be illustrated by a power diagram. Figure 5.11 illustrates the flow of power components in the Kramer cascade, neglecting the losses. Figure 5.11a is a simplified block diagram, where the arrows indicate positive direction. The slip power is recovered from the rotor side, adapted to the grid side in the converter, and merges with the stator power resulting in the cascade electrical power. Figure 5.11b shows simplified Sankey diagrams, relevant for motor operation ($s > 0$) and generator operation ($s < 0$).

The simplified power balance is established under the same assumption of neglecting the losses. The following equations describe active power quantities P and reactive power quantities Q, as for the machine stator side M, the converter W, and the resulting grid values.

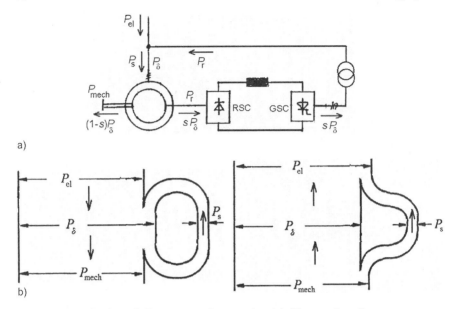

Fig. 5.11 Visualization of Kramer cascade operation (**a**) Diagram for slip power recovery (losses neglected); (**b**) Sankey diagrams; left: motor operation ($s > 0$); right: generator operation ($s < 0$)

$$P_M = P_\delta \quad ; \quad Q_M = P_\delta \cdot \tan \varphi_M$$

$$P_W = -s \cdot P_\delta = -P_s \quad ; \quad Q_W = -s \cdot P_\delta \cdot \tan \alpha_W$$

$$P_{el} = P_M + P_W = (1-s) \cdot P_\delta = P_{mech} \quad ; \quad Q_G = Q_M + Q_W = P_\delta \cdot (\tan \varphi_M + \tan \alpha_W)$$
(5.7)

Note that the air-gap power P_δ transferred from stator to rotor is negative when the slip is negative (supersynchronous). Due to the uncontroled rectifier, active power can only flow from the rotor to the converter; hence the slip power must be positive, $P_s = s \cdot P_\delta > 0$.

The main drawback of this concept is a high reactive power demand, consisting of the machine magnetizing component and the inverter control component. Hence a low overall power factor is obtained which becomes poorer as the speed control ratio is increased. Hence ratios of $n_{max}/n_{min} > 1,5$ are considered inappropriate for practice.

Figure 5.12a shows the principal graph of torque vs. speed (or slip, respectively). The limit of variable speed operation was set to $|s_{0,max}| = 0,5$ by appropriate design of the phase-controlled grid-side converter. Any permissible operation in the quadrants I (motor, subsynchronous) and III (generator, supersynchronous) of Fig. 5.12a can be realized, whereas quadrants II and IV are not possible for the circuit given in Fig. 5.9. Figure 5.12b illustrates the power characteristics of the Kramer system with $k_2 = 0,1$, both in motor und generators mode.

5.3 Systems for Feeding into the Grid

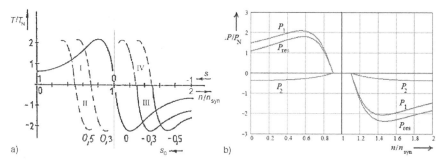

Fig. 5.12 Torque and power characteristics of the cascade system (a) Torque characteristics; (b) Power characteristics, example for $k_2 = 0,1$

5.3.3.3 Doubly-Fed Asynchronous Machine

The term doubly-fed induction machine applies to a system where both stator and rotor winding of a slip-ring machine are supplied. With grid-connected primary winding and a rotor-side converter adapting the slip power to the grid side the system acts as a cascade. Use can be made of a cyclo-converter or a converter with intermediate d.c. circuit as shown in Fig. 5.13. Both rotor-side and grid-side inverter (MSI, GSI) are self-controled devices, allowing active power transfer in both directions and the adjustment of reactive power on both sides. Hence the cascade can, different from the Kramer system in Fig. 5.9, also operate in supersynchronous motor and subsynchronous generator mode. Note that Fig. 5.13 contains a filter (F) and a transformer (T) to adapt the rotor side voltage to the grid. In most cases a crowbar is additionally provided at the rotor-side, by which in case of grid faults the rotor is switched to an external resistor to protect the machine-side inverter from excess current.

For wind energy system purposes the cascade is usually equipped with a VSI type inverter. A common design realizes a speed interval of ratio 1:2, by means of a control such that the range between lower and upper speed is approximately $2/3 < n/n_{syn} < 4/3$, where n_{syn} is the machine synchronous speed at grid frequency.

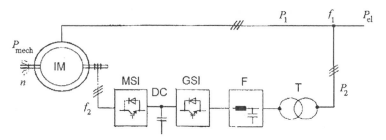

Fig. 5.13 Doubly-fed induction generator with rotor-side converter

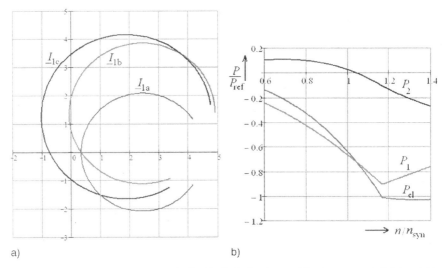

Fig. 5.14 Steady-state performance of a doubly-fed induction machine (a) current loci; (b) variable speed operation (legend see text)

Converter control allows set-value voltage to be fed to the rotor winding, where magnitude and phase can be chosen arbitrarily as long as, for steady state operation, the voltage is of slip frequency. Consider Fig. 5.14a, where per-unit stator current loci appear as circles in the complex plane. The reference circle $\underline{I}_{1,\text{ref}}$ applies to rated stator voltage and rotor short-circuited; it is similar to the circle in Fig. 5.10a. The three circles $\underline{I}_{1a}$, $\underline{I}_{1b}$ and $\underline{I}_{1c}$ were calculated for $k_2 = 0,2$ and $\varphi_2 = \pi$; $1,13\pi$ and $0,885\pi$, respectively, see (5.4). As diameters and center points vary in the examples, the scaling of parameter slip differ also. Note that the current locus $\underline{I}_{1a}$ reflects the circle for the machine with the rotor short-circuited, as in Fig. 5.10a, however with the no-load point at $s = -0,2$. For the other circuits the real current component becomes zero at $s = -0,15$ and $-0,24$, respectively. In the case of $\underline{I}_{1c}$ the rated current in generator operation is reached at power factor unity.

Using a suitable concept, power and power factor of the cascade may be controlled. As an example, consider a performance where, in variable speed operation, the electric power fed into the grid follows a third power law, representing a wind turbine operating at optimum power factor c_p. This applies up to a specified speed from where the power is limited to rated value. Consider a second condition that the stator power factor be unity; this implies that the machine side inverter feeds the magnetizing reactive power into the rotor winding. Fig. 5.14b shows calculation results according to this concept, of stator power P_1, rotor power P_2 and resulting generator power P_{el} expressed in per-unit, for variable speed operation within $-0,4 < s < 0,4$. P_{el} follows a third power function up to the reference point chosen at $n/n_{\text{syn}} = 1,16$, with output limitation above.

5.3.4 Synchronous Generators

5.3.4.1 General

In wind energy application synchronous generators are used in variable speed systems only. The converter to decouple the frequency between machine and grid has to be designed for full load, different from the doubly-fed induction generator concept.

Generators of larger ratings are generally equipped with an excitation winding, fed via slip-rings from a separate exciter. Machines with permanent magnet (PM) excitation have been the domain of smaller ratings, but are gaining their share for ratings also in the Megawatt class.

Synchronous machines are more suitable for designs with large pole numbers than induction machines are. Hence they are the option for direct driven generators, sparing the gear box in the system. Generators of considerable diameter and pole number values are found in the gearless Magawatt systems.

A possible fully-fed system configuration is shown in Fig. 5.15. The synchronous generator SM driven by a wind turbine supplies an uncontrolled rectifier RI to feed a d.c. intermediate circuit DC1 of variable voltage. Connection to a constant voltage intermediate circuit DC2 is made by a step-up inverter SI. The self-controled grid-side inverter GSI is suitable to adjust the power factor. A filter F is used to comply with power quality requirements.

The generators are normally constructed conventionally with interior rotor and radial magnetic flux direction. It is observed that synchronous machines with PM excitation gain more and more interest.

In recent years several unconventional concepts were developed, allegedly offering advantages because of possible integration of generator rotor and wind rotor hub. Solutions avoiding both the heavy gearbox with speed ratios around 100:1 (as in the standard doubly-fed induction generator systems) and the large diameter generator with pole numbers of more than 70 (as in the gearless synchronous machine system) have been introduced. Notable are

– VENSYS concept [Vensys GmbH] using a direct driven permanent magnet exterior rotor synchronous generator,

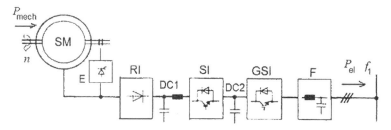

Fig. 5.15 Synchronous generator with full-load converter (legend see text)

- MULTIBRID concept [Multibrid Entwicklungsgesellschaft] using a one-stage gear for speed ratios approximately 10, and
- LIBERTY concept [Clipper Windpower] where a special gear drives four synchronous machines sharing the power production.

Axial flux generators, though frequently proposed for wind energy systems, have only rarely found practical application in the kW range and prototypes of higher rating.

5.3.4.2 Machines with Excitation Winding

In synchronous machines the magnetic flux is provided by the inductor, generally the rotor. Standard generators are provided with a field winding in the rotor, for feeding the d.c. excitation current via slip-rings and brushes. The excitation power is supplied by the exciter which may be a d.c. generator coupled with the main machine. Mostly a static exciter is used which is fed from the generator terminals, as indicated by exciter E in Fig. 5.15. Self-controled generators of smaller ratings for island grids are frequently equipped with compound excitation where the field current is derived from both terminal current and voltage, their values combined vectorially to take phase displacement of the load into account. To start up, self-excitation can be effected by the remanent machine flux.

Slipringless excitation, well-known from turbine generators, is also applicable to other synchronous machines. Fig. 5.16 shows the principle circuitry. The main

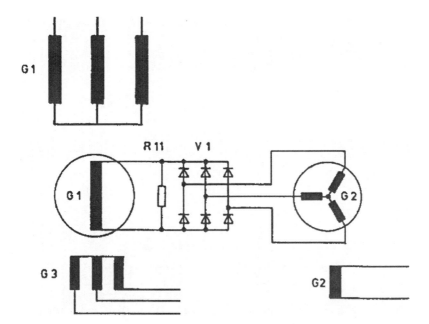

Fig. 5.16 Synchronous generator with slipringless excitation

5.3 Systems for Feeding into the Grid   97

generator G1 has a field winding which is supplied from a coupled a.c. exciter generator G2 having a rotating three-phase armature winding and a stationary field winding. The a.c. excitation power is rectified by means of rotating diodes in a bridge circuit V1. The resistor R11 is a simple protective device on the d.c. side.

A classical extension is the self-controled compound excited generator, where a further auxiliary a.c. winding G3 provides a voltage dependent component which is vectorially added to an armature current component in order to create approximately the required main machine field current. A small electronic controller may be used for improved voltage control.

5.3.4.3 Machines with Permanent Magnet Excitation

Permanent magnet machines feature higher efficiencies than machines with excitation windings (absence of field winding losses), less weight and the advantage of having no slip-rings and brushes. Machines above kilowatt range (and most below) employ high-specific energy density PM material, preferably of Neodymium-iron-boron (NdFeB). Though prices have steadily gone down, the cost of PM material constitutes a considerable part of overall machine cost. A challenge in construction is the installation of the permanent magnets on the rotor, which are mostly mounted polewise in magnetized state on iron supporting elements, for mounting in axial direction. Occuring magnet forces reaching extraordinary high values require rigid auxiliary constructions at the manufacturers.

The material properties of permanent magnets are given in the demagnetization curve; characterized by the remanence flux density B_r, die coercitive force H_c and the maximum remanent energy density $(B \cdot H)_{max}$. Magnetic polarization $J(H)$ and Magnetization $M(H)$ are connected with flux density $B(H)$ by:

$$B = \mu_0 \cdot H + J = \mu_0 \cdot (H + M) \tag{5.8}$$

The coercitive field strength is given by $_BH_C$ (at $B = 0$) or $_JH_C$ (at $J = 0$): here $H_C = {}_BH_C$ is used.

Typical curves $B(H)$ of PM materials are shown in Fig. 5.17 [VAC]. Ferrite magnets are of low specific energy values, but cheap when mass produced; they are still standard in small d.c. motors for consumer or automotive auxiliary drives. AlNiCo magnets, though of high remanence, suffer from low coercitive force values and are currently restricted to very small motors. The properties of rare earth magnets have made remarkable progress in the last years, and they are widely applied in present PM machines. They come as Samarium-Cobalt magnets, especially used in servo motors, and in Neodymium-iron-boron magnets which are favourites, among other applications, for wind turbine synchronous generators.

An example of a NdFeB magnet material, suitable for machine excitation, is VACODYM 655 AP: its catalogue magnetization properties are shown in Fig. 5.18. Typical values are $B_r = 1,2\,\mathrm{T}$, $H_c = 915\,\mathrm{kA/m}$, $(BH)_{max} = 275\,\mathrm{kJ/m^3}$. The dependence on temperature is evident, with the special phenomenon of the "knee"

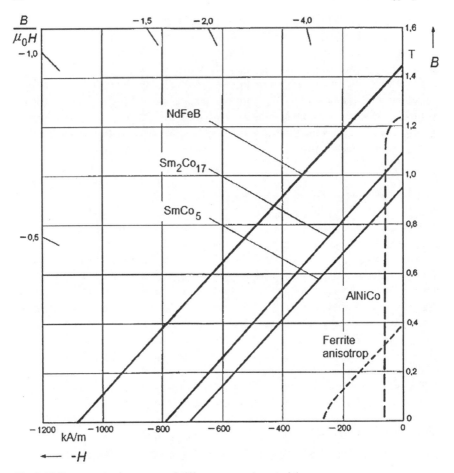

Fig. 5.17 Demagnetization curves of different magnetic materials

appearing in the $B(H)$ and $J(H)$ curves, respectively. Obviously the coercitive field strength absolute value is decreasing with increasing temperature. Machine design must take care that the "knee" is not touched at any load or short-circuit, regarding magnet temperature and armature current reaction (producing m.m.f. in demagnetizing direction). Otherwise the well-known irreversible magnet demagnetization would take place, with a decreased flux and machine capability as consequence.

In case of the material Fig. 5.18 the manufacturer specifies a maximum temperature of 160 °C for application. This is a rather low value, regarding modern machines e.g. of temperature class F. Hence the magnet chosen for example may serve for small machines of moderate utilization only. For wind generators in the MW-class magnets of higher temperature stability are required; the choice may then be the quality VACODYM 677 AP [VAC], with properties a little lower at

5.3 Systems for Feeding into the Grid

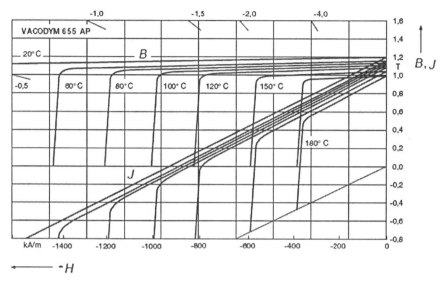

Fig. 5.18 Demagnetization curves of aNdFeB magnets material [VAC]

$B_r = 1,13\,\text{T}$, $H_c = 860\,\text{kA/m}$, $(BH)_{\max} = 240\,\text{kJ/m}^3$, but a maximum temperature of $200\,°\text{C}$ specified for application.

An example of a PM machine is illustrated in Fig. 5.19 where part (a) is the quadrant II representation of the $B(H)$ characteristic; note the demagnetization curve at $20\,°\text{C}$ and $120\,°\text{C}$ and the magnetic circuit characteristic drawn as straight shearing lines for no-load and tracking the armature reaction at short-circuit into account. Part (b) of the figure shows the calculated field pattern of a machine with eight poles; note the magnet supports for 2 half-poles each [Stie00].

It is understood that the performance of permanent magnet synchronous machines is characterized by constant excitation, due to the fixed pole flux value.

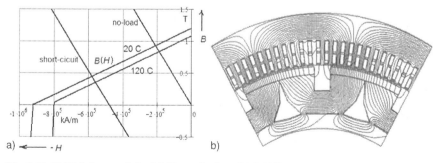

Fig. 5.19 PMSM characteristics (a) Magnetization; (b) field line pattern

Table 5.1 Properties of selected wind energy systems

Manufacturer	Fuhrländer AG	NORDEX AG	Siemens Wind Power GmbH	GE Energy	
Type	FL 250	N60	1,3 MW/62	1.5s	
Rated power	50/250	250/1300	250/1300	1.500	kW
Rated/cut-in/cut-off-wind velocity	15,0/2,5/25,0	15,0/2,5/25,0	13-14/3-4/25	13,0/4,0/25,0	m/s
Rotor					
Diameter/swept area	29,0/661	60,0/2828	62.0/3019	70.5/3.904	m/m^2
No. of blades	3	3	3	3	
Rotational speed	29/38	15,3/23.8	13/19	11 … 20	min^{-1}
Gear box					
stages/ratio	2/1:26	3/1:78	3/1:78	3/1:90	
Generator					
Type	IM, pole change.	IM, pole change	IM, pole change	IM, doubly fed	
speed	750/1000	1010/1515	1000/1500	1500 … 1800	min^{-1}
Voltage	400	690	690	690	V
Grid paralleling method	Thyristors	Thyristors	Thyristors	inverter	
Control/Safety					
Power limitation	Stall	stall	active-stall	pitch	
Main brake	Blade tip var.	Blade tip var.	Blade angle var.	Blade angle var.	
2nd brake	Disk brake	Disk brake	Disk brake	Blade angle var.	
Masses					
Rotor	4,9	21,5	30.0	28,0	t
Nacelle except rotor	9,8	51.4	50,0	49,0	t
Tower					
Hub height	42/50	46/60/69	68/80/90	55/65/80/85	m
Reference yield					
Energy yield		2342	2746	3326	MWh/a
at height of tower		60	68	65	m

a) Example 250 kW, and 1300–1500 kW class

Table 5.1 (continued)

Manufacturer	ENERCON GmbH	Siemens Wind Power GmbH	Nordex AG	
Type	E-70	2.3 MW/82-VS	N90	
Rated power	2300	2300	2300	kW
Rated/cut-in/cut-off-wind velocity	12*/2,5*/28...34	13-14/3-4/25	13/3/25	m/s
Rotor				
Diameter/swept area	71,0/3959	82,4/5333	90	m/m^2
No. of blades	3	3	3	
Rotational speed	6...21,5	6–18	9,6–16,9	min^{-1}
Gear box				
stages/ratio	—	3/1:91	3/1:77,4	
Generator				
Type	SM	IM, fully fed	IN, doubly fed	
speed	6...21,5	550...1600	740...1310	min^{-1}
Voltage	400	690	660	V
Grid paralleling method	Converter	converter	converter	
Control/Safety				
Power limitation	Pitch	pitch	pitch	
Main brake	Blade angle var.	Blade angle var.	Blade angle var	
2nd brake	Blade angle var.	Disk brake	Disk brake	
Masses				
Rotor	31,7*	54	53	t
Nacelle except rotor	68,8*	82	91	t
Tower				
Hub height	64/85/99/113,5	58,5/90/100	80/100/105	m
Reference yield				
Energy yield	4814		5734	MWh/a
at height of tower	85		80	m

b) 2300 kW class

Table 5.1 (continued)

Manufacturer	GE Energy	Vestas Deutschland GmbH	Winwind Oy	
Type	GE 3.0s	V90-3,0 MW	WWD-§	
Rated power	3000	3000	3000	kW
Rated/cut-in/cut-off-wind velocity	14/3/25	16/4/25	12,5/3/20–25	m/s
Rotor				
Diameter/swept area	100/7854	90/6362	100/7854	m/m^2
No. of blades	3	3	3	
Rotational speed	5–18,7	9–19	5–15	min^{-1}
Gear box				
stages/ratio	3/1:107,9	3/1:104,5	*/*	
Generator				
Type	SM (PM)	IM, doubly fed	SM (PM)	
speed	1813 n_N	1680 n_N	*	min^{-1}
Voltage	690	1000	*	V
Grid paralleling method	Converter	thyristors	Converter	
Control/Safety				
Power limitation	Pitch	pitch	Pitch	
Main brake	Blade angle var.	Blade angle var.	Blade angle var.	
2nd brake	Blade angle var.	Disk brake	Disk brake	
Masses				
Rotor	49	28	*	t
Nacelle except rotor	85	68	*	t
Tower				
Hub height	70	80/105	90/100	m
Reference yield				
Energy yield				MWh/a
at height of tower				m

c) 3000 kW class

5.3.5 Examples of Commercial Systems

Table 5.1 presents a compilation of characteristics of different sizes and concepts of commercially available wind energy systems, selected at random from a market survey as examples [Source: Windenergie 2006 –Market survey-, BWE]. Several classes of rated power are covered by the three technical concepts currently in use:

– induction generators, with two speeds by pole-changing (5.3.2),
– induction generators, doubly-fed, with rotor-side converter (5.3.3.3),
– synchronous generators, fully-fed, with converter (5.3.4),
  partly with permanent magnet excitation (5.3.4.3)

5.4 Systems for Island Operation

5.4.1 Systems in Combined Generation

5.4.1.1 General

The main drawback of wind power converters is the unsteady performance dependent on wind velocity. Especially below cut-in speed no power is available. Therefore in island grid application other power sources and/or power storage devices are necessary to satisfy consumer requests.

A well-known combination consists of Diesel generation and wind systems. There are several concepts aimed to serve different load types.

In fuel-saving operation one or more Diesels are continually running; the wind system is connected when there is sufficient wind. The Diesel generator is preferably a self-controled synchronous machine. The wind generator may be a direct coupled induction machine, its magnetizing current being supplied by the Diesel generator.

When there is enough wind to serve the consumer needs, the Diesel can be stopped. A free-wheel clutch allows operating the synchronous machine as a rotating condenser parallel to the induction generator.

Short-time storage devices in stand-alone systems are mostly lead-acid batteries. Cost aspects normally prevent their expansion to long-time storage equipment.

Note that a switching concept depending on consumer priorities can also be considered, where lighting and communication equipment have first priority, and household appliances and heating have to stand back in times of low wind velocity.

5.4.1.2 Combination with Diesel Generation

Figure 5.20 is the sketch of an island grid configuration, containing two Diesels with synchronus generators, and rotating phase shifter for controlled supply of reactive power. Two small and one medium sized wind systems with induction generators

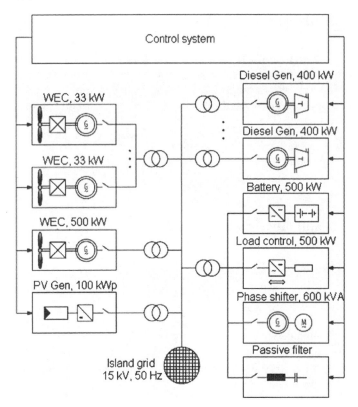

Fig. 5.20 Concept of an island grid with combined generation and storage (Source: SMA)

contribute to power generation. Two capacitor banks are for the basic reactive load compensation. A battery storage is provided which is coupled to the grid via a converter The battery may be combined with the load controller. A passive filter completes the generation system.

The behaviour of simple concepts using frequency as controled variable is shown in Fig. 5.21. The aim is to utilize the wind power wherever possible and to reduce the operation time of the Diesels and save fuel. In the figure the congtroled frequency interval is $49 < f < 51$ Hz which is acceptable only in smaller stand-alone systems. The operation management has to deal with frequent starts, synchronizations and shut-downs. A priority switching scheme for consumers can be incorporated in the system.

5.4.1.3 Combination with Other Renewable Power Sources

An extension of an island system as in Fig. 5.21, without Diesel generation but with another renewable energy source such as photovoltaic (PV) generator, is shown in

5.4 Systems for Island Operation

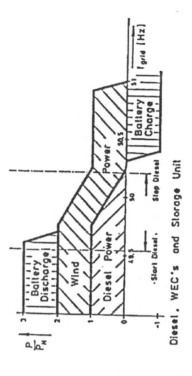

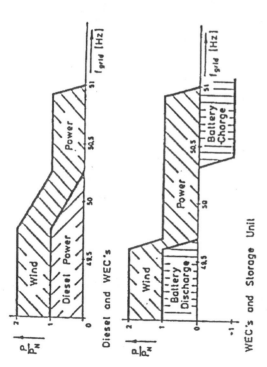

Fig. 5.21 Variants of frequency control

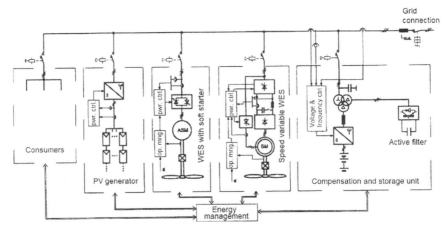

Fig. 5.22 Concept of an autonomous island system with renewable energy generation

Fig. 5.22. Two wind energy generators, one with a synchronous machine, the other with an induction machine, are combined with a solar generating system. Central part of the system is a management and compensation unit which the authors of [Sou01] called an "electronic synchronous machine" because of its ability to control active and reactive power. To bridge short-time peak loads and wind weakness, e.g. for 10 min, a battery storage is provided which is coupled with the grid via a self-controled converter. The device is responsible for frequency and voltage control. It is completed by an active filter to reduce harmonics.

5.4.2 Stand-Alone Systems

5.4.2.1 Small Systems in the kW Range

The market for wind energy systems (WES) with electric output in stand-alone configuration is currently growing. Often a combination with other regenerative sources such as photovoltaic systems (PVS) is intended. The range of ratings considered is from below 1 kW up to around 20 kW. Low investment cost is a criterion especially in this area. Small turbines are normally equipped with fixed blades. Effort and cost for maintenance should be low, in view of suitability for rural or less developed areas. Safe operation properties is indispensable, considering operation by unskilled persons in residential areas.

Most stand-alone systems are for low-voltage ac output. They are mostly equipped with an energy storage device, generally a lead-acid battery of suitable capacity. The design of small size wind systems should realize:

– low cut-in speed,
– simple speed control to run with best-point power coefficient up to rated power,

5.4 Systems for Island Operation

- passive azimuth adjustment,
- power limitation by a preferably passive acting mechanism.

Permanent magnet excited synchrous machines (PMSM) are the preferred generators, due to their "grid-building" properties. More or less elaborated inverters are selected for voltage and frequency adaption. Inverters may be adapted from photovoltaic systems. Asynchronous generators are less often used for small stand-alone systems.

5.4.2.2 Systems with PMSM

Low power wind energy systems of some 100 W up to some kW are frequently used to feed a buffer battery, with a dc or inverter ac output to consumers in an island supply. In these systems called battery loaders the electrical power is generally generated by a direct driven PMSM. The speed dependent generator voltage has to be adapted to the battery voltage.

Figure 5.23 shows an example of measured characteristics of a small battery loader with a rotor diameter of 1,5 m and a rated output power of 400 W at 900 min^{-1}. Note that loading starts at a speed of roughly 300 min^{-1}. Compared with Fig. 3.18 (passive load) the curves are shifted to begin at the cut-in speed, due to friction and electrical loss.

To adapt the ac generator output to the dc battery side a load controller is needed to obtain a satisfactory load characteristic. A simple solution is using a diode rectifier and a step-down dc-dc inverter. Considering systems without pitch-control measures have to be taken to limit the battery voltage in times of weak or no consumer load and to limit the speed at wind velocities above rated value. Power limitation may be effected by mechanical or electrical provisions. Options on the mechanical side are decreasing the cross section of area A with respect to the wind direction

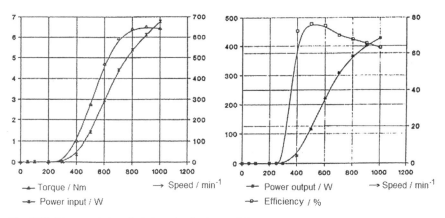

Fig. 5.23 Characteristics of a battery loader (example)

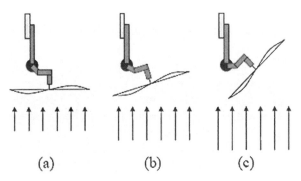

Fig. 5.24 Sketch of furling action

by a furling or tilting mechanism, or to provide or passive operated pitching of the blades. On the electrical side a dump load resistor activated at a speed threshold is a possible solution. Besides, with PMSM as generators electric braking is possible feature.

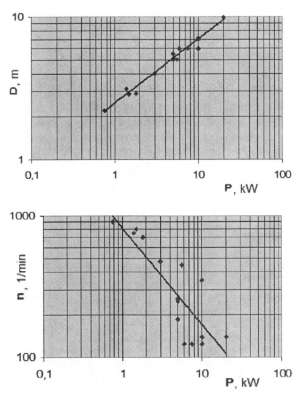

Fig. 5.25 Turbine diameters and maximum speed of commercially available small WES

5.4 Systems for Island Operation

The principle of furling is illustrated in Fig. 5.24, where (a) is for normal wind speeds (no furling); (b) indicates starting of furling action and (c) is for the end position at high wind speeds [Wha05].

For systems of ratings 1 kW and above, available market data allow a view on practical values of turbine diameter and upper operational speed in dependence of power. Figure 5.25 covers the range up to 20 kW; see also [BWE07].

Inverter circuits to couple generator and dc (and battery) side may be selected from the types indicated in Fig. 5.26 where (a) stands for the active front-end inverter, (b) shows a diode rectifier, (c) is the rectifier and step-up inverter combination, and (d) is a reduced version of (c) applicable for use with a generator of high inductance [Wha05].

A comparison can be deduced from Table 5.2: The features determining the cost is the semiconductor count and the complexity of control equipment. It is seen that (a) is the most advanced but costly solution; (b) will not meet requirements of a VSI circuit with controlled dc side voltage; (c) has the advantage of using only 1 switch (transistor) and a simple control; (d) is only suitable with generators of unconventional design.

For systems rated some kW the rectifier and step-up inverter combination (indicated by c) in Fig. 5.26 and Table 5.2 is a good compromise regarding cost and performance.

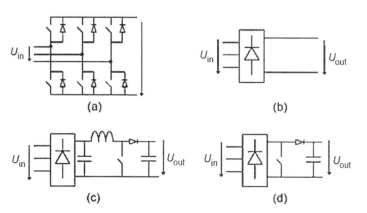

Fig. 5.26 Typical inverter circuits for small ratings

Table 5.2 Comparison of inverter circuits in Fig. 5.26

Inverter type	Diode count	Switch count	Control	Relative cost
a	6	6	complex	high
b	6	0	none	low
c	7	1	simple	medium
d	7	1	simple	medium

Fig. 5.27 Circuit of a system with step-up inverter

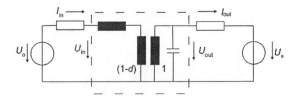

A simplified equivalent circuit is illustrated in Fig. 5.27, where the generator is represented by the induced no-load voltage U_o, is proportional to speed, and an internal resistance. The step-up converter is modelled by an ideal transformer with ratio $1/(1-d)$, an inductor in the input branch and a capacitor in the output branch. The load side consists of a resistor and an e.m.f. representing the constant battery voltage U_B.

The performance of an example system for 4 kW at 350 min^{-1} is illustrated by calculated curves in Fig. 5.28. Characteristics are shown for battery loading operation, under the condition of continuous current. The duty ratio d is controlled to keep torque proportional to speed, between 80 and 350 min^{-1}. In the Figure part (a) shows the course of the generator voltage vs. speed and input and output voltage of the step-up inverter; part (b) depicts input and output currents, while (c) is the duty ratio; eventually (d) is the course of input power.

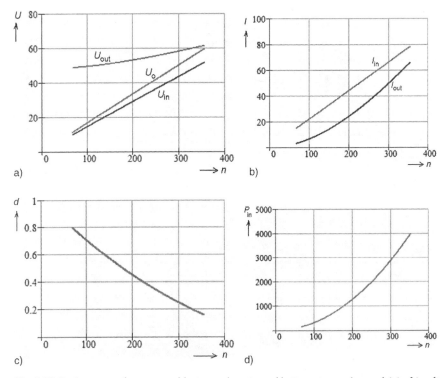

Fig. 5.28 Performance of a system with step-up inverter and battery storage (example) (**a–b**) voltages, currents; (**c**) duty ratio; (**d**) input power

5.4 Systems for Island Operation

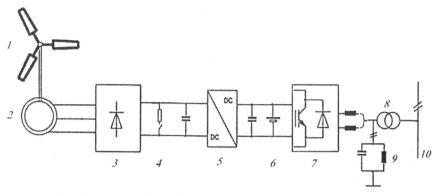

Fig. 5.29 Stand-alone system with synchronous generator and battery storage

The concept of a stand-alone system in the kW-range is illustrated in Fig. 5.29. The synchronous, three-phase PM excited generator 2 is driven by the wind turbine *1*. A first d.c., variable voltage intermediate circuit is fed via the diode bridge rectifier *3*. A step-up inverter (boost chopper) *5* supplies the second d.c. circuit containing a storage battery *6*. An active front-end inverter *7* feeds the single-phase local grid *10*. In practice a transformer *8* will be necessary to match battery and grid voltage; *9* is a passive filter for harmonics reduction, and *4* indicates a short-circuiting device. Note similarities to the circuit in Fig. 5.15.

5.4.2.3 Systems with Induction Generator

The self-excited induction generator concept (SEIG) with capacitors as reactive power source has often been recommended in literature. Simple solutions with constant or switched capacitor banks are, however, not suitable for practical use due to the load and speed dependence of the terminal voltage; see Fig. 3.12. An acceptable solution would require appropriate voltage and reactive power control.

A simple possibility is to use a phase-controlled inductance (see Fig. 3.10a) together with a capacitor bank. A pilot set-up is shown in Fig. 5.30, where the output voltage is controlled by means of a thyristor inverter, adjusting the phase angle α. The passive LR-load is connected to the generator system by closing the switch S.

More sophisticated SEIG control concepts use active filters [chat06].

The supply of magnetizing current to a cage induction machine used as a generator can be achieved by means of an active front end inverter, see 4.3.3.3. Figure 5.31 indicates a stand-alone system, where C_1 is designed to provide a basic compensation to take some of the inverter load, but not sufficient for creating self-excitation. The scheme contains a buck-boost dc-dc inverter with an inductive storage element; hence the polarity from left to right side is reversed. This solution with two intermediate dc-circuits is suitable to cover a wide speed range, however at considerable

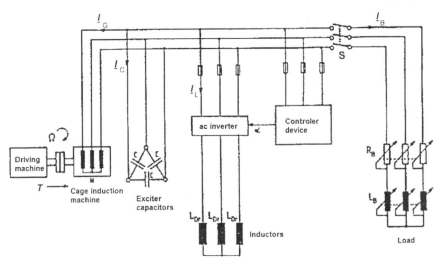

Fig. 5.30 Self-excited induction generator with phase control

cost and increased losses. The grid-side inverter is a self-controled device, which allows to control the power factor. A single-phase output is assumed. Further shown is a passive filter to keep harmonics off the output, and a transformer to adapt the voltage levels of battery and grid. Note that the normal machine operation is supplying active power and consuming reactive power, i.e. in the III. quadrant.

For an induction machine used as generator in a stand-alone system, two control concepts may be used:

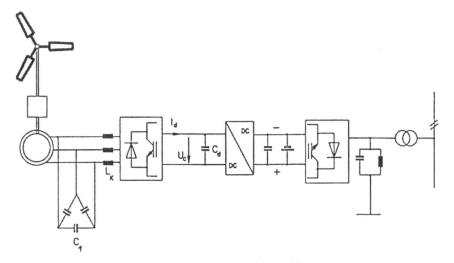

Fig. 5.31 Stand-alone system with cage induction machine and battery storage

- Operation at constant flux. The induced voltage is roughly proportional to the speed, and maximum generator power is obtained at rated current.
- Operation at maximum efficiency. This is the case when constant losses and load dependent losses are equal. In this mode the relation between voltage and speed is non-linear. Note that standard induction machines are designed to have maximum efficiency around 80% of rated load.

Normally the constant flux concept is preferred, because of larger energy output.

Chapter 6
Performance and Operation Management

6.1 General

Performance of the generators has been dealt with (Chap. 3) in steady state condition, covering operation of three-phase machines on symmetrical supply of constant voltage, constant frequency and at constant rotational speed. Investigating transient phenomena, unsymmetrical conditions, stability and behavior under control, however, requires extended machine models, for which different methods are available.

Performance of power electronic devices was described (Chap. 4) for phase and switch-mode controlled inverters also in steady-state condition, i.e. without input and load disturbances, describing average operation characteristics. Adapted models are required to model transient operation conditions.

This chapter describes suitable methods for system component modeling and gives an introduction to system control and operation management.

6.2 System Component Models

6.2.1 Model Representation

6.2.1.1 Modal Component Transformations

To investigate operation conditions and transient phenomena in three-phase a.c. systems modal components are widely used [IEC62428]. Where inductive, resistive or capacitive couplings between phase elements and lines occur, they are useful to simplify description and calculation of these phenomena. The original time-dependent quantities are transformed into modal components, to give decoupled impedances and admittances.

Modal components are used for a.c. power systems, including electric machines. Basics of definitions and methods are given in this chapter.

Original quantities g_1, g_2, g_3 and the modal components $\underline{g}_{M1}$, $\underline{g}_{M2}$, $\underline{g}_{M3}$ are related to each other by the transformation equation:

M. Stiebler, *Wind Energy Systems for Electric Power Generation*. Green Energy and Technology, © Springer-Verlag Berlin Heidelberg 2008

$$\begin{bmatrix} g_1 \\ g_2 \\ g_3 \end{bmatrix} = \begin{bmatrix} \underline{t}_{11} & \underline{t}_{12} & \underline{t}_{13} \\ \underline{t}_{21} & \underline{t}_{22} & \underline{t}_{23} \\ \underline{t}_{31} & \underline{t}_{32} & \underline{t}_{33} \end{bmatrix} \begin{bmatrix} \underline{g}_{M1} \\ \underline{g}_{M2} \\ \underline{g}_{M3} \end{bmatrix} \tag{6.1}$$

or in shortened form

$$g = \underline{T} \cdot \underline{g}_M$$

The coefficients t_{tk} of the transformation matrix can be real or complex. It is necessary that the transformation matrix $\underline{T}$ is non-singular, so that the inverse relationship is valid:

$$g_M = \underline{T}^{-1} \cdot g$$

If the relationship between the original quantities and the modal components is introduced for voltages as well as for currents, then:

$$u = \underline{T}\, \underline{u}_M; \quad i = \underline{T}\, i_M$$

Transformations mainly used in the field of electrical machines are the Clarke ($\alpha\beta0$-), the Park (dq0) and the space-phasor transformation.

6.2.1.2 Power-Invariant and Power-Variant Transformation

The power p expressed in terms of the original quantities is:

$$p = u_1\, i_1^* + u_2\, i_2^* + u_3\, i_3^* = (u_1\ u_2\ u_3) \begin{bmatrix} i_1^* \\ i_2^* \\ i_3^* \end{bmatrix} = u^T i^* \tag{6.2}$$

where i^* denotes the conjugate complex value of i.

In terms of modal components the instantaneous power is expressed by:

$$p = \underline{u}_M^T\, (\underline{T}^T\, \underline{T}^*)\, i_M^* \tag{6.3}$$

The transformation is power-invariant, if $\underline{T}$ is chosen so that $(\underline{T}^T\, \underline{T}^*) = E$, where E is the unity matrix; consequently the power-independent transformation matrix $\underline{T}$ is unitary, so that

$$\underline{T}^{-1} = (T^T)^* \tag{6.4}$$

On the other hand, in the power-variant forms of transformation, also known as reference-component-invariant transformations, the properties of $\underline{T}$ are such that under balanced symmetrical conditions the reference component of the modal components is equal to the reference component of the original quantities. In case of the Clarke transformation this means $(\underline{T}^T\, \underline{T}^*) = 3/2 \cdot E$, see Table 6.1.

6.2 System Component Models   117

The following paragraphs describe the main transformations used for investigating electrical machine performance, in power-invariant form.

6.2.1.3 Transformation into $\alpha\beta 0$-Components (Clarke Transformation)

Original components g_a, g_b, g_c are transformed to orthogonal components g_α, g_β, g_0 by using the real transformation matrix $\underline{T}_\alpha$:

$$\underline{T}_\alpha = \sqrt{\frac{2}{3}} \begin{bmatrix} 1 & 0 & 1/\sqrt{2} \\ -1/2 & \sqrt{3}/2 & 1/\sqrt{2} \\ -1/2 & -\sqrt{3}/2 & 1/\sqrt{2} \end{bmatrix} \tag{6.5}$$

The original components of the three-phase system are mapped by two components of a two-phase system, plus a zero-sequence component.

The $\alpha\beta 0$-transformation is usually applied for calculations in stator reference frame.

6.2.1.4 Transformation into dq0-Components (Park Transformation)

Original components g_a, g_b, g_c are transformed to orthogonal components g_d, g_q, g_0 by using the real transformation matrix $\underline{T}_d$:

$$\underline{T}_d = \sqrt{\frac{2}{3}} \begin{bmatrix} c_1 & -s_1 & 1/\sqrt{2} \\ c_2 & -s_2 & 1/\sqrt{2} \\ c_3 & -s_3 & 1/\sqrt{2} \end{bmatrix} \tag{6.6}$$

where

$$c_1 = \cos\varphi; \quad c_2 = \cos(\varphi - 2\pi/3); \quad c_1 = \cos(\varphi + 2\pi/3)$$
$$s_1 = \sin\varphi; \quad s_2 = \sin(\varphi - 2\pi/3); \quad s_1 = \sin(\varphi + 2\pi/3)$$

Similar as above, 2 components and the zero-sequence component are created; however the coordinate system is turned by angle φ which for a rotating machine is a time-dependent variable.

The dq0-transformation is usually applied for calculations in rotating reference frame, especially when the reference frame is fixed to the rotor.

6.2.1.5 Transformation into Space Phasors

Space phasors or space vectors are complex quantities representing a two-phase system; plus the zero-sequence component required to ensure a reversible transformation algorithm. For real original components the properties are such that the second component is the conjugate complex of the first.

Original components g_a, g_b, g_c are transformed to components g_s, $g_z = g_s^*$, g_0 by using the complex transformation matrix $\underline{T}_s$:

For a non-rotating frame and with unchanged reference axis (direction a = real axis), the transformation is defined by:

$$\underline{T}_s = \frac{1}{\sqrt{3}} \begin{bmatrix} 1 & 1 & 1 \\ \underline{a}^2 & \underline{a} & 1 \\ \underline{a} & \underline{a}^2 & 1 \end{bmatrix} \tag{6.7}$$

where $\underline{a} = e^{j2\pi/3}$; $\underline{a}^2 = \underline{a}^*$; $1 + \underline{a} + \underline{a}^2 = 0$

For a rotating frame, where the coordinate system is turned by an angle ϑ, the transformation is defined by:

$$\underline{T}_s = \frac{1}{\sqrt{3}} \begin{bmatrix} e^{j\vartheta} & e^{-j\vartheta} & 1 \\ \underline{a}^2 e^{j\vartheta} & \underline{a} e^{-j\vartheta} & 1 \\ \underline{a} e^{j\vartheta} & \underline{a}^2 e^{-j\vartheta} & 1 \end{bmatrix} \tag{6.8}$$

Note that in practice ϑ is often chosen to define a field vector as reference direction; the value of the reference vector is then a real. On the other hand, the reference frame is fixed to the rotor of a machine rotating at instantaneous velocity Ω by choosing:

$$\vartheta = \int \Omega(t) dt \tag{6.9}$$

6.2.1.6 Transformation into Symmetrical Components (Fortescue Transformation)

The transformation transforms complex original quantities into complex symmetrical components $\underline{g}_{(1)}$, $\underline{g}_{(2)}$, $\underline{g}_{(0)}$. The transformation matrix is the same as (6.6), applicable to map original phasors of a three-phase system by positive-sequence, negative-sequence and zero-sequence components.

This transformation is mentioned for completeness; its application is normally for unsymmetrical states of distribution grids.

6.2.1.7 Transformations and Reference Frames

In practice the transformation matrices are often used in power-variant form. The matrices for $\alpha\beta 0$-transformation, dq0-transformation, sz0-transformation with non-rotating and rotating frame and their respective inverse are given in Tables 6.1

6.2 System Component Models

Table 6.1 Transformation matrices in the power-variant form I

	$\times [i_\alpha\ i_\beta\ i_0]^T$	$\times [i_d\ i_q\ i_0]^T$	$\times [i_s\ i_z\ i_0]^T_{nr}$	$\times [i_s\ i_z\ i_0]^T_r$
$\begin{bmatrix} i_a \\ i_b \\ i_c \end{bmatrix} =$	$\begin{bmatrix} 1 & 0 & 1 \\ -1/2 & \sqrt{3}/2 & 1 \\ -1/2 & -\sqrt{3}/2 & 1 \end{bmatrix}$	$\begin{bmatrix} c_1 & -s_1 & 1 \\ c_2 & -s_2 & 1 \\ c_3 & -s_3 & 1 \end{bmatrix}$	$\dfrac{1}{2}\begin{bmatrix} 1 & 1 & 2 \\ \underline{a}^2 & \underline{a} & 2 \\ \underline{a} & \underline{a}^2 & 2 \end{bmatrix}$	$\dfrac{1}{2}\begin{bmatrix} e^{j\vartheta} & e^{-j\vartheta} & 2 \\ \underline{a}^2 e^{j\vartheta} & \underline{a}e^{-j\vartheta} & 2 \\ \underline{a}e^{j\vartheta} & \underline{a}^2 e^{-j\vartheta} & 2 \end{bmatrix}$
		$c,\ s$ are as in (6.6)		ϑ is e.g. as in (6.9)

Table 6.2 Transformation matrices in the power variant form II

	$\times [i_a\ i_b\ i_c]^T$
$\begin{bmatrix} i_\alpha \\ i_\beta \\ i_0 \end{bmatrix} =$	$\dfrac{2}{3}\begin{bmatrix} 1 & -1/2 & -1/2 \\ 0 & \sqrt{3}/2 & -\sqrt{3}/2 \\ 1/2 & 1/2 & 1/2 \end{bmatrix}$
$\begin{bmatrix} i_d \\ i_q \\ i_0 \end{bmatrix} =$	$\dfrac{2}{3}\begin{bmatrix} c_1 & c_2 & c_3 \\ -s_1 & -s_2 & -s_3 \\ 1/2 & 1/2 & 1/2 \end{bmatrix}$
$\begin{bmatrix} i_s \\ i_z \\ i_0 \end{bmatrix}_{nr} =$	$\dfrac{2}{3}\begin{bmatrix} 1 & \underline{a} & \underline{a}^2 \\ 1 & \underline{a}^2 & \underline{a} \\ 1/2 & 1/2 & 1/2 \end{bmatrix}$
$\begin{bmatrix} i_s \\ i_z \\ i_0 \end{bmatrix}_{r} =$	$\dfrac{2}{3}\begin{bmatrix} e^{-j\vartheta} & \underline{a}\,e^{-j\vartheta} & \underline{a}^2 e^{-j\vartheta} \\ e^{j\vartheta} & \underline{a}^2 e^{j\vartheta} & \underline{a}e^{j\vartheta} \\ 1/2 & 1/2 & 1/2 \end{bmatrix}$

and 6.2 in the power-variant version. The tables indicate how original components are deduced from modal components, and vice versa, for the currents as example. Transformation matrices to connect different modal systems may be deduced from the Tables; they are listed in [IEC62428].

Note that the matrices regarding $\alpha\beta0$- and dq0-transformation consist of real elements. The two-phase components may, however, be combined in complex form, e.g. $\underline{g} = g_\alpha + j\,g_\beta$.

In the power-variant form (6.4) is no longer valid, and specific constant factors have to be applied for power calculation from modal components.

Figure 6.1 illustrates the application to AC machines in general. Figure 6.1a models a cross-section of a two-pole, three-phase machine with a smooth air-gap between stator and rotor. The stator carries the windings a, b, c; with $120°$ displacement between any two of them. On the rotor also a three-phase system with the windings k, l, m is assumed. In the figure the rotor position angle is γ, the electrical angle between the reference axes a of the stator and k of the rotor.

Figure 6.1b shows the coordinates as assigned to the modal components defined above. The Clarke transformation (6.4) assigns stator components α, β to the original components a, b, c. The reference axis of the coordinate system remains the

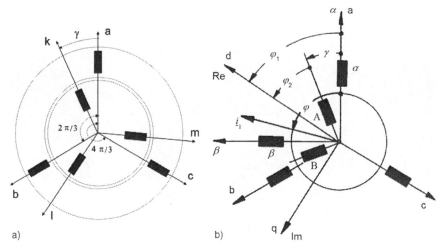

Fig. 6.1 Three-phase machine model and coordinates (**a**) three-phase windings in stator and rotor; (**b**) different frames

same (direction a = α). Regarding the rotor, transformed components are A, B, assigned to original components k, l, m. The zero-sequence component is not shown in the figure; it may be drawn separately because according to the machine model it does not contribute to torque production.

Park's transformation (6.5) assigns components d, q to the original components. Reference axis d may be chosen arbitrarily. Using a common coordinate system for stator and rotor quantities, φ in the transformation according to (6.5) will be $\varphi = \varphi_1$ for stator and $\varphi = \varphi_2$ for rotor quantities, see Fig. 6.1. In synchronous machines the coordinate system is generally fixed to the rotor, with the pole-axis as reference. When the d, q components are expressed by a complex quantity, $\underline{g}_s = g_\alpha + j\, g_\beta$, the d-axis is the real axis.

It may be mentioned that the use of modal components are not restricted to three-phase a.c. machines, but finds wide application in the analysis of power distribution networks.

6.2.2 Asynchronous Machine Models

6.2.2.1 Model in $\alpha\beta$-Coordinates

The machine to be considered carries a symmetrical three-phase winding in the stator. The rotor may be a wound rotor with slip-rings, or a cage rotor. For purpose of the model, the rotor impedance is considered independent of slip (i.e. no current displacement taken into account), the air-gap is constant and magnetic saturation effects are neglected. Skin-effect in rotor conductors and iron saturation may be taken into account in a refined model later.

Describing the electrical system of asynchronous machines, the windings are represented in stator and rotor by two orthogonal windings on each member. Figure 6.2

6.2 System Component Models

Fig. 6.2 Induction machine model in Clarke components

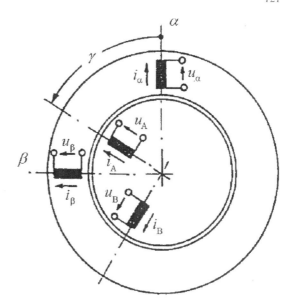

shows the model in $\alpha\beta$-components. It is assumed that the three-phase windings are in Y-connection, and also that no neutral conductor is connected, so that the zero-sequence component need not be considered.

The $\alpha\beta 0$ transformation on the original stator components *abc* will be performed in the power-variant form according to Tables 6.1 and 6.2. The same matrix is applied to voltages and currents, so that the transformation equation and its inverse is given by:

$$\underline{u}_{abc} = \underline{T}_\alpha \cdot \underline{u}_{\alpha\beta 0} \quad ; \quad \underline{i}_{abc} = \underline{T}_\alpha \cdot \underline{i}_{\alpha\beta 0}.$$

$$\underline{u}_{\alpha\beta 0} = \underline{T}_\alpha^{-1} \cdot \underline{u}_{abc} \quad ; \quad \underline{i}_{\alpha\beta 0} = \underline{T}_\alpha^{-1} \cdot \underline{i}_{abc} \qquad (6.10)$$

In any moment the power in the three-phase system

$$P_{abc} = \underline{u}_{abc}^T \underline{i}_{abc} = \underline{u}_{\alpha\beta 0}^T (\underline{T}_\alpha^T \cdot \underline{T}_\alpha) \underline{i}_{\alpha\beta 0} \qquad (6.11)$$

As mentioned before, in the power-variant transformation $\underline{T}_\alpha$ is not a unitary matrix. The matrix $(\underline{T}_\alpha^T \cdot \underline{T}_\alpha)$ is diagonal and constant; for a system without zero-sequence current it indicates a multiplying factor 3/2.

Same as this orthogonal transformation for the stator quantities, the rotor quantities for an assumed symmetrical three-phase winding system transform the original components *klm* into modal components *AB0*, where the reference coordinate axis A is rotated by the angle γ with respect to the stationary reference axis α. In the following, zero-sequence components will not be considered.

The voltage equation of the four-winding model, not taking zero-sequence components into account, and assuming short circuited rotor winding, equals the applied stator terminal voltage to the sum of ohmic and inductive voltage drops; capacitive voltage drops are negligible for modelling the low frequency behaviour. The inductive components are described by the time-derivative of the flux linkages.

$$
\begin{bmatrix} u_\alpha \\ u_\beta \\ 0 \\ 0 \end{bmatrix} = \begin{bmatrix} R_s & 0 & 0 & 0 \\ 0 & R_s & 0 & 0 \\ 0 & 0 & R_r & 0 \\ 0 & 0 & 0 & R_r \end{bmatrix} \begin{bmatrix} i_\alpha \\ i_\beta \\ i_A \\ i_B \end{bmatrix} + \frac{d}{dt} \begin{bmatrix} \psi_\alpha \\ \psi_\beta \\ \psi_A \\ \psi_B \end{bmatrix} \tag{6.12}
$$

where

$$
\begin{bmatrix} \psi_\alpha \\ \psi_\beta \\ \psi_A \\ \psi_B \end{bmatrix} = \begin{bmatrix} L_s & 0 & L_m\cos\gamma & -L_m\sin\gamma \\ 0 & L_s & L_m\sin\gamma & L_m\cos\gamma \\ L_m\cos\gamma & L_m\sin\gamma & L_r & 0 \\ -L_m\sin\gamma & L_m\cos\gamma & 0 & L_r \end{bmatrix} \begin{bmatrix} i_\alpha \\ i_\beta \\ i_A \\ i_B \end{bmatrix}
$$

The flux linkages ψ depend on the winding currents i. The coefficient matrix contains the self-inductances L_s, L_r, and the magnetizing inductance L_m. Conventionally, the self-inductances are the sum of magnetizing inductance and leakage components not coupled with any other winding:

$$
L_s = L_m + L_{\sigma s} \quad ; \quad L_r = L_m + L_{\sigma r}.
$$

The mutual inductances between stator and rotor consist of L_m multiplied by trigonometric functions of the rotor position angle γ.

When expressing stator and rotor quantities by space phasors $\underline{g}_s = (g_\alpha + \mathrm{j}\, g_\beta)$ and $\underline{g}_r = (g_A + \mathrm{j}\, g_B)$, the voltage equations assume a shortened form:

$$
\begin{bmatrix} \underline{u}_1 \\ 0 \end{bmatrix} = \begin{bmatrix} R_s & 0 \\ 0 & R_r \end{bmatrix} \begin{bmatrix} \underline{i}_1 \\ \underline{i}_2 \end{bmatrix} + \frac{d}{dt} \begin{bmatrix} \underline{\psi}_1 \\ \underline{\psi}_2 \end{bmatrix} \tag{6.13}
$$

where

$$
\begin{bmatrix} \underline{\psi}_1 \\ \underline{\psi}_2 \end{bmatrix} = \begin{bmatrix} L_s & L_m \cdot e^{\mathrm{j}\gamma} \\ L_m \cdot e^{-\mathrm{j}\gamma} & L_r \end{bmatrix} \begin{bmatrix} \underline{i}_1 \\ \underline{i}_2 \end{bmatrix}
$$

The electromagnetic torque, also called the air-gap torque is calculated from known space phasors:

$$
T_{el} = \frac{3}{2} z_p L_m \cdot \mathrm{Im}\left[\underline{i}_1 \underline{i}_2^* \cdot e^{-\mathrm{j}\gamma}\right] = \frac{3}{2} z_p \cdot \mathrm{Im}[\underline{i}_1 \underline{\psi}_1^*] \tag{6.14}
$$

Assuming a rotating system of one inertia J, the equation of rotation equals the acceleration torque with the sum of air-gap torque and load torque:

$$
\frac{J}{z_p} \frac{d^2\gamma}{dt^2} = T_{el} + T_L \tag{6.15}
$$

6.2 System Component Models

Accelerating (motoring) torque values are counted as positive. Comparison of the moment of inertia of different sizes of systems is simplified when defining the unit acceleration time, the time which would be required to bring the rotating parts of a machine from rest to rated speed if the accelerating torque were constant and equal to the quotient of rated active power by rated angular velocity:

$$\tau_j = \frac{J\Omega_N^2}{P_N} = \frac{J}{P_N}\left(\frac{\omega_{sN}}{z_p}\right)^2 \tag{6.16}$$

6.2.2.2 Modal Component Model

Now a new coordinate system is defined in which stator and rotor quantities are referred to a single, rotating reference axis, with the advantage of eliminating the angular functions in (6.12) and (6.13). This may be done by using the transformation:

$$\begin{bmatrix} i_1 \\ i_2 \end{bmatrix} = \begin{bmatrix} e^{j\varphi_1} & 0 \\ 0 & e^{j(\varphi_1-\gamma)} \end{bmatrix} \begin{bmatrix} i_s \\ i_r \end{bmatrix} \quad \text{where} \quad \frac{d\varphi_1}{dt} = \omega_{ref} \; : \; \frac{d\gamma}{dt} = \omega_{rot} = z_p\Omega \tag{6.17}$$

Note that ω_{rot} is the mechanical angular velocity of the rotor, multiplied by the number of pole pairs. In the example of Fig. 6.1 the reference is the d-axis, also the real axis for complex quantities.

Applying this transformation to (6.13), in which ω_{ref} is an arbitrary angular speed of the reference coordinate system and ω the rotational speed, we get:

$$\begin{bmatrix} \underline{u}_s \\ 0 \end{bmatrix} = \begin{bmatrix} R_s & 0 \\ 0 & R_r \end{bmatrix} \begin{bmatrix} \underline{i}_s \\ \underline{i}_r \end{bmatrix} + \frac{d}{dt}\begin{bmatrix} \underline{\psi}_s \\ \underline{\psi}_r \end{bmatrix} + j\begin{bmatrix} \omega_{ref} & 0 \\ 0 & (\omega_{ref}-\omega_{rot}) \end{bmatrix}\begin{bmatrix} \underline{\psi}_s \\ \underline{\psi}_r \end{bmatrix} \tag{6.18}$$

where

$$\begin{bmatrix} \underline{\psi}_s \\ \underline{\psi}_r \end{bmatrix} = \begin{bmatrix} L_s & L_m \\ L_m & L_r \end{bmatrix} \begin{bmatrix} \underline{i}_s \\ \underline{i}_r \end{bmatrix}$$

The relevant equivalent circuit model is shown in Fig. 6.3; note the inductive voltage drop components of transformer type $(d\psi/dt)$ and rotary type $(j\omega\psi)$.

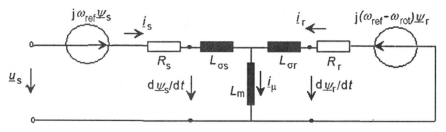

Fig. 6.3 Dynamic induction machine model

124 6 Performance and Operation Management

Table 6.3 Properties of preferred transformations

Coordinate system	φ_1	φ_2	ω_{rel}	$(\omega_{rel} - \omega_{rot})$
fixed to the stator	0	$-\gamma$	0	$-\omega_{rot}$
fixed to the rotor	γ	0	ω_{rot}	0
fixed to synchronous vector	$2\pi ft$	$(2\pi ft - \gamma)$	$2\pi f$	$s \cdot 2\pi f$

In practice, only a limited number of transformations according to (6.17) are of practical interest, see Table 6.3.

Additionally, in drive technology transformations are used where the reference axis is fixed to stator or rotor flux vector, see (6.21). Note that the special case of steady-state operation of the induction machine is included in the equations.

When the machine is fed by a symmetrical three-phase voltage of frequency $\omega_s = 2\pi ft$, the voltage vector in $\alpha\beta0$ components fixed to stator is in the power-variant form:

$$\begin{bmatrix} u_a(t) \\ u_b(t) \\ u_c(t) \end{bmatrix} = \sqrt{2}U_s \begin{bmatrix} \cos(\omega_s t) \\ \cos(\omega_s t - 2\pi/3) \\ \cos(\omega_s t + 2\pi/3) \end{bmatrix} \longrightarrow \begin{bmatrix} u_\alpha \\ u_\beta \\ u_0 \end{bmatrix}$$

$$= \sqrt{2}U_s \begin{bmatrix} \cos(\omega_s t) \\ \sin(\omega_s t) \\ 0 \end{bmatrix} \longrightarrow \underline{u}_s = \sqrt{2}U_s \cdot e^{j\omega_s t} \qquad (6.19a)$$

In an arbitrary coordinate system rotating at an angular frequency ω_{ref} the transformed voltage is, also in power-variant form:

$$\begin{bmatrix} u_d \\ u_q \\ u_0 \end{bmatrix} = \sqrt{2}\,U_s \begin{bmatrix} \cos[(\omega_s - \omega_{ref})t] \\ \sin[(\omega_s - \omega_{ref})t] \\ 0 \end{bmatrix} \longrightarrow \underline{u}_s = \sqrt{2}\,U_s \cdot e^{j(\omega_s - \omega_{ref})t} \qquad (6.19b)$$

Note that in synchronous rotating frame, $\omega_{ref} = \omega_s$, the voltage vector is a constant: $\underline{u}_s = \sqrt{2}U_s$. In steady state operation, the derivatives of fluxes in the model (6.18) vanish, $d\psi/dt = 0$. In this case the model is a proper representation of the steady-state model described in Sect. 3.2.2.

6.2.2.3 Model in Field-Oriented Components

In induction machine drive concepts often a flux is chosen as a controlled variable, either the stator or the rotor flux. In field-oriented control this is preferably the rotor flux. Considering a machine controlled to run with impressed stator current, a first-order differential equation for the rotor flux as state variable is obtained:

$$\frac{L_s}{\tau_{0r}}\underline{i}_s = \left[\frac{1}{\tau_{0r}} + j\left(\omega_{ref} - \omega_{rot}\right)\right]\underline{\psi}_r + \frac{d\underline{\psi}_r}{dt} \quad \text{where} \quad T_{0r} = \frac{L_r}{R_r} = \frac{T_{kr}}{\sigma} \qquad (6.20)$$

6.2 System Component Models

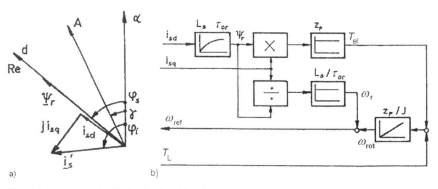

Fig. 6.4 Rotor model of induction machine (**a**) Space-vector diagram; (**b**) block diagram

Choosing the rotor flux space vector as reference, it is positioned in the d-axis which is also the real axis. Hence (6.20) expressed by components becomes:

$$L_s i_{sd} = \psi_r + \tau_{0r} \frac{d\psi_r}{dt} \quad \text{where} \quad \underline{\psi}_r = \psi_{rd} = \psi_r$$
$$L_s i_{sq} = \tau_{0r} \omega_r \psi_r \quad \text{where} \quad \omega_r = \omega_{ref} - \omega_{rot} \tag{6.21}$$

The electromagnetic torque is expressed by:

$$T_{el} = \frac{3}{2} z_p i_{sq} \psi_r \tag{6.22}$$

It is seen that the torque is defined by the product of q-axis current and d-axis flux; consequently i_q is called the torque building component and i_d the flux-building component. The equation of motion for the one-mass inertia model is:

$$J \frac{d\omega_{rot}}{dt} = T_{el} + T_L \tag{6.23}$$

Figure 6.4 shows space vector diagram and a block diagram representation of the rotor model based on (6.20–6.22). In the figure the rotational speed is calculated according to (6.23) which of course may be replaced when a measured speed value is available.

In Fig. 6.4b i_{sd}, i_{sq} and T_L are the input values; calculated results are ψ_r, ω_r and T_{el}. To apply the rotor model, the current components may be obtained from measured currents in stator frame by a transformation, when the rotor angular frequency is known from measurement or estimation. For flux oriented control, rotor flux and torque signals serve as the actual values.

6.2.2.4 Transient Model

In order to calculate the behavior of a cage induction machine subjected to variations of the load torque a transient model can be used. Consider a machine operating at a constant symmetrical voltage, described in synchronous frame with constant voltage. The stator circuit will be assumed in steady state, i.e. transients are neglected. When neglecting also the stator resistance, the system is simplified to an algebraic stator equation and a (complex) first-order differential equation for the rotor:

$$\underline{U}_s = j\omega_s \underline{\psi}_s \quad ; \quad [1+j\tau_{kr}\omega_r]\underline{\psi}_r + \tau_{kr}\frac{d\underline{\psi}_r}{dt} = \frac{L_m}{L_s}\underline{\psi}_s \qquad (6.24a)$$

In dq components, with $\underline{U}_s = U_s$, we have:

$$\tau_{kr} \cdot \frac{d}{dt}\begin{bmatrix} \psi_{rd} \\ \psi_{rq} \end{bmatrix} + \begin{bmatrix} 1 & -\tau_{kr}\omega_r \\ \tau_{kr}\omega_r & 1 \end{bmatrix}\begin{bmatrix} \psi_{rd} \\ \psi_{rq} \end{bmatrix} = \frac{L_m}{L_s}\frac{\sqrt{2}\,U_s}{\omega_{ref}}\begin{bmatrix} 0 \\ -1 \end{bmatrix} \qquad (6.24b)$$

The electric torque is expressed equivalent to (6.14) as

$$T_{el} = -\frac{3}{2}z_p\frac{L_m}{L_s}\frac{\sqrt{2}\,U_s}{\omega_s\sigma L_r}\cdot\psi_{rd} \qquad (6.25)$$

Together with the equation of motion (6.23) for a single inertia drive a second-order system is obtained describing the transient machine model.

6.2.2.5 Model for Small Deviations from Steady State

In a number of problems, e.g. forced oscillations due pulsating load torque, the quantities may be described by small deviations from a steady-state operation. A linear system in terms of deviations of flux $\Delta\underline{\psi}_r$ and rotor pulsation $\Delta\omega_r$ from a steady state characterized by $\underline{\psi}_{r0}$ and ω_{r0} (or s_0) may then be established:

$$[1+j\tau_{kr}\omega_{r0}]\Delta\underline{\psi}_r + \tau_{kr}\frac{d\Delta\underline{\psi}_r}{dt} = -j\tau_{kr}\underline{\psi}_{r0}\Delta\omega_r \quad \text{where} \quad \underline{\psi}_{r0} = -j\frac{\sqrt{2}U_s}{\omega_s}\frac{1}{1+j\omega_{r0}\tau_{kr}}$$
$$(6.26)$$

The system is completed by the equation of motion which is, on the basis of (6.22, 6.23) for small deviations:

$$-\tau_J P_N z_p\frac{d\Delta\omega_r}{dt} = \Delta T_{el} + \Delta T_L \quad \text{where} \quad \Delta T_{el} = -\frac{3}{2}z_p\frac{L_m}{L_s}\frac{\sqrt{2}U_s}{\omega_s\sigma L_\gamma}\text{Re}(\Delta\underline{\psi}_r) \quad (6.27)$$

A simple solution of the linear system is obtained for steady state at no-load, $\omega_{r0} = 0$. Using the notation of Laplace transform with p = differential operator, the rotor frequency deviation as output with respect to a small load torque deviation is described by the response:

$$\frac{\Delta\omega_r}{\Delta T_L} = \frac{\omega_s}{T_N K_D}\frac{1+p\tau_{kr}}{1+p\tau_J/K_D+p^2\tau_{kr}\tau_J/K_D} \qquad (6.28)$$

where

$$K_D = \frac{2T_k}{T_N} \omega_{ref} \tau_{kr}$$

is the initial slope of the per-unit torque characteristic vs. slip,

$$T_N = \frac{z_p P_N}{\omega_N} \qquad \text{is the rated torque.}$$

Note that the breakdown-slip is $s_k = 1/(\omega_{ref} \tau_{kr})$. The relevant characteristic equation is:

$$p^2 \Delta \omega_r + p2\delta \, \Delta \omega_r + v_0^2 \Delta \omega_r = 0 \qquad (6.29)$$

where

$$\delta = 1/(2\tau_{kr})$$

and

$$v_0^2 = \frac{K_D}{\tau_{kr} \tau_J}$$

For $v_0^2 > \delta^2$ the solution is conjugate complex:

$$p_{1,2} = -\delta \pm j\sqrt{v_0^2 - \delta^2} = -\delta \pm j\omega_e$$

This case is of interest since in most practical the machine, reacting on a disturbance, oscillates at an electromechanical eigenfrequency. This is observed e.g. during run-up of grid-supplied induction machines.

Example: With parameters $T_k/T_N = 2,5$ and $s_k = 0,1$ the damped eigenfrequency is $f_e = \omega_e/(2\pi) = 6,3$ Hz at $\tau_J = 1$ s and $f_e = 1,3$ Hz at $\tau_J = 5$ s.

6.2.3 Synchronous Machine Models

6.2.3.1 Machine with Field Winding on the Rotor

The synchronous machine representation in Fig. 6.5a contains a stator with the three-phase windings a, b, c, and a rotor, generally of the salient pole type, carrying a field winding f. The stator quantities may be transformed, in stationary frame, into α, β components, see Fig. 6.5b. Park's transformation leads to the d, q components, the reference d-direction fixed to the rotor direct (magnetizing) axis. The Figure does not take care of zero-sequence components. The model

The voltage equations in the three-windings model α, β, f for the synchronous machine without damper circuits in the rotor is:

$$\begin{bmatrix} u_\alpha \\ u_\beta \\ u_f \end{bmatrix} = \begin{bmatrix} R_s & 0 & 0 \\ 0 & R_s & 0 \\ 0 & 0 & R_f \end{bmatrix} \begin{bmatrix} i_\alpha \\ i_\beta \\ i_f \end{bmatrix} + \frac{d}{dt} \begin{bmatrix} \psi_\alpha \\ \psi_\beta \\ \psi_f \end{bmatrix} \qquad (6.30)$$

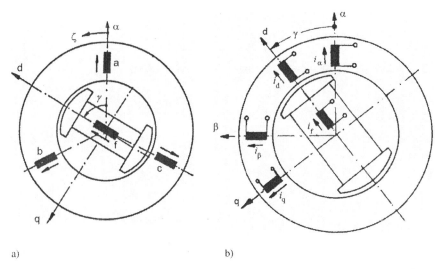

Fig. 6.5 Synchronous machine model (**a**) Salient pole three-phase machine; (**b**) Transformed windings arrangement

where

$$\begin{bmatrix} \psi_\alpha \\ \psi_\beta \\ \psi_f \end{bmatrix} = \begin{bmatrix} (L_d \cos^2\gamma + L_q \sin^2\gamma) & (L_d - L_q)\sin\gamma\cos\gamma & L_{md}\cos\gamma \\ (L_d - L_q)\sin\gamma\cos\gamma & (L_d \sin^2\gamma + L_q \cos n^2\gamma) & L_{md}\sin\gamma \\ L_{md}\cos\gamma & L_{md}\sin\gamma & L_f \end{bmatrix} \begin{bmatrix} i_\alpha \\ i_\beta \\ i_f \end{bmatrix}$$

The electromagnetic torque becomes:

$$T_{el} = \frac{3}{2} z_p (i_\beta \psi_\alpha - i_\alpha \psi_\beta) \tag{6.31}$$

The equation of motion of the single-inertia model is (6.15), same as for the asynchronous machine.

The transformation into Park components follows the definition:

$$\begin{bmatrix} i_\alpha \\ i_\beta \end{bmatrix} = \begin{bmatrix} \cos\gamma & -\sin\gamma \\ \sin\gamma & \cos\gamma \end{bmatrix} \begin{bmatrix} i_d \\ i_q \end{bmatrix} \leftrightarrow \begin{bmatrix} i_d \\ i_q \end{bmatrix} = \begin{bmatrix} \cos\gamma & \sin\gamma \\ -\sin\gamma & \cos\gamma \end{bmatrix} \begin{bmatrix} i_\alpha \\ i_\beta \end{bmatrix} \tag{6.32}$$

Using the definition for quantities i, u and ψ, the following voltage equation is obtained:

$$\begin{bmatrix} u_d \\ u_q \\ u_f \end{bmatrix} = \begin{bmatrix} R_s & 0 & 0 \\ 0 & R_s & 0 \\ 0 & 0 & R_f \end{bmatrix} \begin{bmatrix} i_d \\ i_q \\ i_f \end{bmatrix} + \frac{d}{dt}\begin{bmatrix} \psi_d \\ \psi_q \\ \psi_f \end{bmatrix} + \frac{d\gamma}{dt}\begin{bmatrix} -\psi_q \\ \psi_d \\ 0 \end{bmatrix} \tag{6.33}$$

where

$$\begin{bmatrix} \psi_d \\ \psi_f \end{bmatrix} = \begin{bmatrix} L_d & L_{md} \\ L_{md} & L_f \end{bmatrix} \begin{bmatrix} i_d \\ i_f \end{bmatrix} \quad ; \quad \psi_d = L_q i_q$$

6.2 System Component Models

In case of permanent magnet (PM) excited machines the notion of an impressed exciter current is used, and

$$\psi_p = L_{md} i_f$$

is defined as the inductor pole flux, which is obviously a constant for a given machine.

Now the expression for the torque becomes:

$$T_{el} = \frac{3}{2} z_p \left(i_q \psi_d - i_d \psi_q \right) \tag{6.34}$$

In grid operation, when the machine is fed from a symmetrical system of frequency $\omega_s = 2\pi f$, the voltage component representation in α, β and d, q components is described by (6.19), where the definition contains the load angle ϑ, which in steady-state operation is a constant, due to the then synchronous rotor speed.

$$\begin{bmatrix} u_\alpha \\ u_\beta \end{bmatrix} = \sqrt{2} U_s \cdot \begin{bmatrix} \cos(\omega_s t) \\ \sin(\omega_s t) \end{bmatrix} \rightarrow \begin{bmatrix} u_d \\ u_q \end{bmatrix} = \sqrt{2} U_s \cdot \begin{bmatrix} \sin \vartheta \\ \cos \vartheta \end{bmatrix}; \quad \text{where } \vartheta = \gamma - \omega_s t + \frac{\pi}{2} \tag{6.35}$$

6.2.3.2 Transient Model

A suitable means to calculate several non-steady regimes of a synchronous machine is the transient model. The approach is to consider steady state stator equations, which means neglecting transients decaying with short-circuit time constants. A further simplification is made by neglecting the stator resistance voltage drop, giving:

$$\begin{bmatrix} U_d \\ U_q \end{bmatrix} = \omega_s \begin{bmatrix} -\psi_q \\ \psi_d \end{bmatrix}$$

In the remaining non-linear first order rotor differential equation the transient inductor e.m.f., usually called e'_p, is chosen for state variable:

$$\tau'_d \frac{de'_p}{dt} + e'_p \frac{X'_d}{X_d} u_p + \left(1 - \frac{X'_d}{X_d} \right) \cdot U_s \cos \vartheta; \tag{6.36}$$

where

$$u_p = u_f \frac{X_{md}}{R_f} \quad ; \quad e'_p = \omega_s \frac{X_{md}}{X_f} \psi_f$$

and X'_d, τ'_d are as defined with (6.47).

The parameters are now, besides the direct axis synchronous reactance $X_d = \omega_s L_d$, the direct axis transient reactance $X'_d = \omega_s L'_d$ and direct axis transient time-constant τ'_d. The electromagnetic torque becomes:

$$T_{el} = -\frac{3 z_p}{2 \omega_s} \left[\frac{U_s \cdot e'_p}{X'_d} \cdot \sin \vartheta - \frac{U_s^2}{2} \left(\frac{1}{X'_d} - \frac{1}{X_q} \right) \sin(2\vartheta) \right] \tag{6.37}$$

The equation of motion describes the behavior of the rotor due to variations in load torque T_L and/or excitation voltage u_p. In the following equation $\dot{\vartheta}$ is the velocity describing the deviation from the steady-state load angle. A damping torque is assumed to be proportional to $\dot{\vartheta}$ with the constant parameter K_D, representing friction and windage losses, and partly rotor losses due to rotor eddy-currents.

$$\frac{J}{z_p}\frac{d\dot{\vartheta}}{dt} + K_D\dot{\vartheta} = T_{el} + T_L \quad \text{where} \quad \dot{\vartheta} = \frac{d\vartheta}{dt} = z_p\Omega - \omega_s \tag{6.38}$$

Equations (6.25) to (6.27) form a non-linear second-order differential equation system. J denotes the resulting moment of inertia of the whole rotating system, calculated by taking gear transmissions into account.

For generator operation at constant terminal voltage, the model allows to determine oscillations due to changes in load torque and/or excitation voltage. When a constant e_p' is assumed, the load angle $\vartheta(t)$ is the only state variable.

6.2.3.3 Model for Small Deviations from Steady State

Linearization of the system (6.26), (6.27) for small deviations $\Delta\vartheta$ from a steady-state operation point, with deviations ΔT as the input variable leads to a second-order system

$$\frac{\tau_J}{\omega_s}\Delta\ddot{\vartheta} + \frac{K_D\omega_s}{P_N}\Delta\dot{\vartheta} + \frac{T_{s0}\omega_s}{P_N}\Delta\vartheta = \frac{\omega_s}{P_N}\Delta T_L \tag{6.39}$$

where τ_J is the unit acceleration time according to (6.16), and T_{s0} is the synchronizing torque at the operation point characterized by the steady-state load angle ϑ_0:

$$T_{s0} = \left(\frac{dT_{el}}{d\vartheta}\right)_{\vartheta=\vartheta_0} = \frac{3z_p}{2\omega_s}\left[\frac{U_s \cdot e_p'}{X_d'}\cdot\cos\vartheta_0 - U_s^2\left(\frac{1}{X_d'} - \frac{1}{X_q}\right)\cos(2\vartheta_0)\right] \tag{6.40}$$

The steady-state load angle may be determined for a machine at synchronous speed, operating at active power P and reactive power Q, with Q positive when supplying reactive power (over-excitation):

$$\vartheta_0 = -\text{atan}\left(P\left/\left(\frac{3U^2}{2X_q}+Q\right)\right.\right)$$

The linear differential equation may be expressed in the conventional form

$$\Delta\ddot{\vartheta} + 2\delta\,\Delta\dot{\vartheta} + v_0^2\,\Delta\vartheta = \frac{\omega_s^2}{\tau_J P_N}\Delta T_L$$

Its characteristic equation is:

$$p^2\Delta\vartheta + p2\delta\Delta\vartheta + v_0^2\Delta\vartheta = 0 \tag{6.41}$$

6.2 System Component Models

where

$$\delta = K_D \omega_s^2 / (P_N \tau_J)$$

and

$$v_0^2 = \frac{T_{s0} \omega_s^2}{P_N \tau_J}$$

For $v_0^2 > \delta^2$ it has a conjugate complex solution:

$$p_{1,2} = -\delta \pm j \sqrt{v_0^2 - \delta^2} \tag{6.42}$$

Under the condition $v_0^2 > \delta^2$ which is mostly the case, except for drives with large inertia values, the system features an electro-mechanical eigenfrequency which is observed in rotor oscillations. Typical frequency values are a few Hz, their values decreasing with increasing unit acceleration time.

6.2.3.4 Machine with a Damper Cage

In order to take transient rotor currents into account, the model of Fig. 6.5 must be amended. The simplest way is to add one damper (amortisseur) mesh on the rotor in each axis. The equivalent damper windings D, Q are short-circuited. The machine may then be represented by an equivalent circuit model in d,q-components with five windings

The voltage equation is an extension of (6.22) and given by:

$$
\begin{bmatrix} u_d \\ u_q \\ 0 \\ 0 \\ u_f \end{bmatrix}
=
\begin{bmatrix} R_s & & & & \\ & R_s & & 0 & \\ & & R_D & & \\ & & & R_Q & \\ 0 & & & & R_f \end{bmatrix}
\begin{bmatrix} i_d \\ i_q \\ i_D \\ i_Q \\ i_f \end{bmatrix}
+ \frac{d}{dt}
\begin{bmatrix} \psi_d \\ \psi_q \\ \psi_D \\ \psi_Q \\ \psi_f \end{bmatrix}
+ \frac{d\gamma}{dt}
\begin{bmatrix} -\psi_q \\ \psi_d \\ 0 \\ 0 \\ 0 \end{bmatrix}
\tag{6.43}
$$

The flux linkages are expressed with the reactances as parameters:

$$
\begin{bmatrix} \omega \psi_d \\ \omega \psi_D \\ \omega \psi_f \end{bmatrix}
=
\begin{bmatrix} X_d & X_{md} & X_{md} \\ X_{md} & X_D & X_{Df} \\ X_{md} & X_{Df} & X_f \end{bmatrix}
\begin{bmatrix} i_d \\ i_D \\ i_f \end{bmatrix}
\quad : \quad
\begin{bmatrix} \omega \psi_Q \\ \omega \psi_Q \end{bmatrix}
=
\begin{bmatrix} X_q & X_{mq} \\ X_{mq} & X_Q \end{bmatrix}
\begin{bmatrix} i_q \\ i_Q \end{bmatrix}
$$

The reactance components are

$$X_d = X_{md} + X_{\sigma s} \qquad\qquad X_q = X_{mq} + X_{\sigma s}$$
$$X_D = X_{Df} + X_{\sigma D} = X_{md} + X_{rc} + X_{\sigma D} ; \quad X_Q = X_{mq} + X_{\sigma Q}$$
$$X_d = X_{Df} + X_{\sigma f} = X_{md} + X_{rc} + X_{\sigma f}$$

The model in Fig. 6.6, based on a salient pole machine, contains provisions for a magnetic leakage flux coupling of field and direct axis damper windings. The

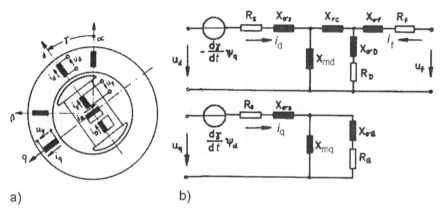

Fig. 6.6 Model of the synchronous machine with five windings (**a**) windings arrangement in d,q components; (**b**) equivalent circuit model

coupling reactance value X_{rc} may be positive (as in turbogenerators) or negative (as in many salient-pole machines). Frequently, the machine model is simplified using $X_{rc} = 0$, resulting in larger errors of calculated rotor quantities.

6.2.3.5 Reactance Operators and Frequency Response

The model of (6.31) and Fig. 6.6 may alternatively be described in terms of transient and subtransient reactances (or inductances) and time-constants. Most of them can be determined by tests.

A well-known model representation is by means of reactance operators, which are complex functions of the differential operator p [IEC60034, part 4]. Using this approach, the voltage equation is:

$$u_d(p) = R_s i_d(p) + p\psi_d(p) - \omega\psi_q(p) \qquad (6.44)$$
$$u_q(p) = R_s i_q(p) + p\psi_q(p) + \omega\psi_d(p)$$

where

$$\omega = \frac{d\gamma}{dt}$$

The flux linkages are in operator representation:

$$\omega\psi_d(p) = \underline{X}_d(p)i_d(p) + \underline{G}_f(p)u_f(p) \ ; \quad \omega\psi_q(p) = \underline{X}_q(p)i_q^i(p) \qquad (6.45)$$

With known flux operators the currents may be calculated:

$$i_d(p) = +\frac{\omega\psi_d(p)}{\underline{X}_d(p)} - \frac{\underline{G}_f(p)}{\underline{X}_d(p)}u_f(p) \ ; \quad i_q(p) = \frac{\omega\psi_q(p)}{\underline{X}_q(p)} \qquad (6.46)$$

6.2 System Component Models

The admittance operators are expressed using reactances and time constants. The conventional form which contains a number of approximations is used here:

$$\frac{1}{\underline{X}_d(p)} = \frac{1}{X_d} \cdot \frac{1+p\tau'_{do}}{1+p\tau'_d} \cdot \frac{1+p\tau''_{do}}{1+p\tau''_d} \quad ; \quad \frac{1}{\underline{X}_q(p)} = \frac{1}{X_q} \cdot \frac{1+p\tau'_{qo}}{1+p\tau''_q} \tag{6.47}$$

$$\underline{G}_f(p) = \frac{\omega\tau'_{md}}{1+p\tau'_{do}} \cdot \frac{1+p\tau_{dD\sigma}}{1+p\tau''_{do}}$$

The coefficients in terms of the parameters of the equivalent circuit model, Fig. 6.6b, are the transient and subtransient reactances

$$X'_d = X_d - \frac{X^2_{md}}{X_f} \quad ; \quad X''_d = X_d - \frac{X^2_{md}(X_D + X_f - 2X_{md})}{X_D X_f - X^2_{md}} \quad \text{and} \quad X''_q = X_d - \frac{X^2_{mq}}{X_Q}$$

and the time constants

$$\tau'_{do} = \frac{X_f}{\omega R_f} \quad ; \quad \tau''_{do} = \frac{X_D - X^2_{md}/X_f}{\omega R_D} \quad ; \quad \tau''_{qo} = \frac{X_Q}{\omega R_Q}$$

$$\tau'_d = \frac{X'_d}{X_d}\tau'_{do} \quad ; \quad \tau''_d = \frac{X''_d}{X'_d}\tau''_{do} \quad ; \quad \tau''_q = \frac{X''_q}{X_q}\tau''_{qo}$$

$$\tau'_{md} = \frac{X_{md}}{\omega R_f} \quad ; \quad \tau_{dD\sigma} = \frac{X_{\sigma D}}{\omega R_D}$$

Additionally, the armature short-circuit time constant is defined:

$$\tau_a = \frac{1}{\omega R_s} \frac{2}{1/X''_d + 1/X''_q}$$

The reactance operators are used for investigations in the Laplace domain or in the frequency domain, e.g. for problems of forced oscillations, $p = j\nu$, or asynchronous operation, $p = js\omega$.

6.2.4 Converter Modeling

Most of the converters used in wind energy systems are voltage source ac-ac inverters (VSI) with intermediate dc circuit, see 4.3.4. While two-level inverters are standard in low-voltage systems, large WES are increasingly designed for medium voltage, e.g. 3kV, where three-level inverters are used in order to limit currents and reduce losses; their modelling is considered in [Ale06]. Besides, dc-dc converters are used, either as step-up or step-down converters, see 4.3.5.

Inverter performance as described in literature is mostly either averaging input-output behaviour or modelling the switching functions in the time domaine. Steady state of an ac.-dc current-source inverter (CSI) is conventionally modelled assuming sinusoidal ac-voltage and constant dc-side current, taking account of harmonics of the uneven order components in the ac side current and even order components in the dc side voltage, see 4.3.2.1. A dual performance is exhibited by the voltage source inverter (VSI), when sinusoidal ac side current and constant dc side voltage is assumed; consequent harmonics appear in the ac voltage and in the dc current, see e.g. 4.3.3.4.

To describe the small-signal behaviour of a semiconductor device in terms of network theory, the hybrid model may be used, as for a transistor, linking input and output sinusoidal quantities:

$$\begin{bmatrix} \underline{U}_1 \\ \underline{I}_2 \end{bmatrix} = \begin{bmatrix} \underline{h}_{11} & \underline{h}_{12} \\ \underline{h}_{21} & \underline{h}_{22} \end{bmatrix} \begin{bmatrix} \underline{I}_1 \\ \underline{U}_2 \end{bmatrix} \qquad (6.48)$$

This equation corresponds with the circuit model in Fig. 6.7, the parameters of which are explained as follows:

$\underline{h}_{11}$ = short-circuit input impedance;
$\underline{h}_{12}$ = open-circuit voltage reaction;
$\underline{h}_{21}$ = short-circuit current gain;
$\underline{h}_{22}$ = open-circuit output admittance.

An intermediate circuit VSI constitutes a time-variant non-linear system. Under certain conditions the model of Fig. 6.7 may be extended to include both fundamentals and dc and ac side harmonics. This requires a sophisticated analysis, such as described in [San06].

In the case of an ac-dc inverter, the ac side input mesh, with terminal voltage $\underline{U}_1$, consists of a ohmic-inductive RL series impedance and voltage source coupled with output voltage $\underline{U}_2$. The output mesh contains a capacitive C admittance and a current source coupled with the input current.

The complete converter is then modelled as in Fig. 6.8 [Ack05]. Let $\underline{U}_1$ be the generator terminal voltage (in case of fully fed system) or the rotor terminal voltage (in case of doubly fed asynchronous generator), and $\underline{U}_2$ the grid side terminal voltage. The intermediate dc circuit is modelled by the two source currents $I_{dc,1}$, $I_{dc,1}$ and the capacitance C.

In most simulations, however, constant dc (capacitor) voltage is assumed, and the inverter is modelled by a transfer factor only, neglecting losses and energy storage properties. Furthermore, only ac quantities of fundamental frequency and constant dc quantities are regarded. In this case, the VSI model is simplified to a real transfer function dependent only on the modulation factor, see 4.3.3.

Dc-dc converters are mostly modelled by the averaging method [Nirg01].

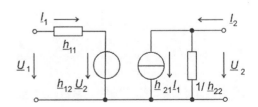

Fig. 6.7 Four-pole equivalent circuit model in hybrid form

6.2 System Component Models

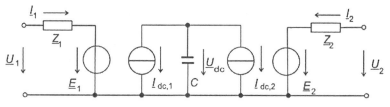

Fig. 6.8 Model of intermediate circuit voltage source converter

6.2.5 Modeling the Drive Train

For investigating transient phenomena and stability problems a model of the mechanical drive train must be established. This is usually done by a lumped parameter model of the rotating components. The one-mass system considered in the previous chapters is not sufficient to study occurring oscillations and torsional stress. The usual approach is to describe the rotating system by a multi-mass model.

A multi-inertia system is described by the differential matrix vector equation:

$$\boldsymbol{J}\ddot{\varphi} + \boldsymbol{D}\dot{\varphi} + \boldsymbol{K}\varphi = \boldsymbol{T} \qquad (6.49)$$

where

φ is the vector of angular rotor positions,
$\boldsymbol{J}$ is the matrix of inertias,
$\boldsymbol{D}$ is a damping matrix,
$\boldsymbol{K}$ is the matrix of spring coefficients, and
$\boldsymbol{T}$ is the vector of torques acting on the inertias.

Note that angular velocities are $\Omega = d\varphi/dt$. In the present problem the inertia components J are arranged in a line, coupled by shaft components characterized by spring parameters K, while damping parameters D are attributed to losses in the inertia components themselves and in the coupling shaft components. The system (6.49) will be considered linear.

Considering a general wind energy system, the main components that make up the rotating inertia are the wind turbine, the gear-box and the generator. Hence it is advisable to model the system by three masses or, when there is no gear-box, by two-masses. Such a representation does not take into account that the blades (mostly 3) have their own degrees of freedom and may contribute differently to the turbine torque. However, extension to a six-mass model is not necessary for shaft oscillation and stability calculations.

Figure 6.9 is a sketch of a three-mass lumped parameter model. In simple representation an inertia is modeled by a homogenous disk of radius R, length L and a material density ρ. On the other hand the spring coefficient of a shaft segment is described by its geometric dimensions and the shear modulus G. Disk mass inertia moment and cylinder shaft stiffness become:

Fig. 6.9 Three-mass drive train model of a wind system T wind turbine; B gear box; G generator

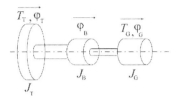

$$J = \frac{\rho}{2}\pi R^4 L \quad ; \quad K = \frac{G}{2}\pi \frac{R^4}{L} \qquad (6.50)$$

Note that the inertia values of two disks in series are added, $J = J_1 + J_2$, while the spring constant of two cylindrical shaft elements in series follows $1/K = 1/K_1 + 1/K_2$. The differential equation assigned to Fig. 6.9 becomes:

$$\begin{bmatrix} J_T & & \\ & J_B & \\ & & J_G \end{bmatrix} \begin{bmatrix} \ddot{\varphi}_T \\ \ddot{\varphi}_B \\ \ddot{\varphi}_G \end{bmatrix} + \begin{bmatrix} D_T & & \\ & D_B & \\ & & D_G \end{bmatrix} \begin{bmatrix} \dot{\varphi}_T \\ \dot{\varphi}_B \\ \dot{\varphi}_G \end{bmatrix} + \begin{bmatrix} K_{TB} & -K_{TB} & \\ -K_{TB} & (K_{TB}+K_{BG}) & -K_{BG} \\ & -K_{BG} & K_{BG} \end{bmatrix} \begin{bmatrix} \varphi_T \\ \varphi_B \\ \varphi_G \end{bmatrix} = \begin{bmatrix} T_T \\ \\ T_G \end{bmatrix}$$

In this model damping effects in the steel shafts and in couplings are neglected, and only viscous damping losses are considered. When a gear-box is present, all quantities have to be referred to either the low-speed side or the high-speed side. Normally it is the latter, i.e. the generator side. The parameters of the turbine side must be referred to the generator side using the speed ratio; as example the equivalent turbine inertia is calculated as:

$$J'_T = \left(\frac{\omega_G}{\omega_T}\right)^2 J_T$$

A parameter example of realistic wind energy systems shows that the relation $J_T : J_B : J_G$ is in the region of 12:0, 6:1. In conclusion a two-mass system may be used as reduced order model. Then the gear box inertia J_B is mostly halved, with the halves added to either side, on the other hand simulation results recommend adding the total gear-box value to the generator.

The two-mass model improves considerably the one-mass representation; it allows for determining one mechanical eigenfrequency and estimation of transient torques stressing shafts and couplings. Therefore its properties are discussed shortly.

Defining the difference $\varphi = \varphi_T - \varphi_G$ as the torsion angle, a second order linear differential equation can be established:

$$\ddot{\varphi} + \underbrace{D\left(\frac{1}{J_T} + \frac{1}{J_G}\right)}_{2dv_0} \dot{\varphi} + \underbrace{K\left(\frac{1}{J_T} + \frac{1}{J_G}\right)}_{v_0^2} \varphi = \frac{T_T}{J_T} + \frac{T_G}{J_G} \qquad (6.51)$$

The two-mass system features a damping coefficient d and a characteristic frequency v_0. Since damping values of the shaft train are low, the response to a step

6.2 System Component Models

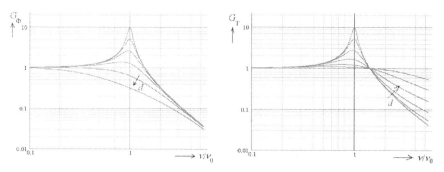

Fig. 6.10 Resonance curves of displacement angle Φ (*left*) and shaft torque T_{sh} (*right*) Damping coefficients $d = 0.05; 0, 1; 0, 2; 0, 4; 0, 8; 1, 6$

input function contains always an oscillation, its frequency increasing with the shaft stiffness, and decreasing with inertia values.

Equation (40) may serve to calculate frequency response curves in case of forced oscillations. Suppose the wind turbine torque contains an oscillating vector component of amplitude $\hat{\underline{T}}_T$ and frequency v, then the resulting torsional angle $\hat{\underline{\Phi}}_T$ and transferred shaft torque $\hat{\underline{T}}_{sh}$, also expressed as vectors, wre described by the equations:

$$\frac{\hat{\underline{\Phi}}}{\hat{\underline{T}}_T} = \frac{1}{K} \cdot K_J \cdot \underline{G}_\Phi \quad : \quad \frac{\hat{\underline{T}}_{sh}}{\hat{\underline{T}}_T} = K_J \cdot \underline{G}_T \tag{6.52}$$

where

$$K_J = \frac{J_G}{J_T + J_G} \quad : \quad \underline{G}_\Phi = \frac{1}{(1 - v^2/v_0^2) + j2d(v/v_0)} \quad ;$$

$$\underline{G}_T = \frac{1 + j2d(v/v_0)}{(1 - v^2/v_0^2) + j2d(v/v_0)}$$

In case of the displacement angle the first factor is the shaft stiffness reversed, characterizing the component as an integrator. Further the equations consist of two factors. The factor K_f relates to the mass forces; it denotes the proportion of a step torque exerted at the turbine side which is led through the shaft and acts on the generator. With a high-inertia of the turbine rotor compared with the generator rotor, K_f is relatively small and the dynamic shaft torque is well below the input torque. This is however the case only when the resonant properties are not taken into account. This factors $\underline{G}$ are the well-known resonant functions of a two-mass system: they appear in principle in same form for electrical resonant circuits. Their magnitudes as functions of the impressed oscillation frequency v are shown in Fig. 6.10.

The curves of G_T indicate the benefit of tuning a system so that a probable disturbance frequency is placed well above the system eigenfrequency: the load side is then even better isolated from oscillations with small damping values than with high damping.

6.3 System Control

6.3.1 General

6.3.1.1 Optimum Control by Model Characteristic

Control concepts operating without measuring the wind speed can be used when wind turbine parameters are known. Considering the variable wind direction and velocity the determination of correct test values is not a simple task. Methods outlined in the following may therefore be of advantage for small up to medium speed-variable systems.

The torque produced by a wind turbine can be described in dependence of the rotor speed by rearranging (2.5) using (2.5) in the equation:

$$T = \frac{P}{\Omega} = \frac{\rho}{2} A \left(\frac{D}{2} \right)^3 \frac{c_p(\lambda)}{\lambda^3} \Omega^2 \tag{6.53}$$

According to Fig. 2.6 a turbine operates at $c_{p,\text{max}}$ when the tip speed ratio is equal to the design value λ_A. When setting $\lambda = \lambda_A$ in (6.53) the torque is proportional to the square of the rotational speed, the turbine operating at maximum power. In (6.54) K is a constant for given turbine parameters:

$$T = \left[\frac{\rho}{2} A \left(\frac{D}{2} \right)^3 \frac{c_p(\lambda_A)}{\lambda_A^3} \right] \Omega^2 = K \cdot \Omega^2 \tag{6.54}$$

Equation (6.54) is then the control rule for the torque set-value in dependence of rotational speed. The algorithm is valid between cut-in speed and rated operation for turbines with fixed blade position, i.e. keeping the pitch angle constant.

6.3.1.2 Optimum Control by MPP Tracking

Another method of optimum control of a variable speed wind turbine is based on the maximum power point (MPP) tracking. The method may be illustrated using an analogy with photovoltaic system control. Figure 6.11 (left) shows in principle the characteristics of a PV-generator: currents I versus voltage U with the parameter solar radiation E; the line of maximum power is indicated. In Fig. 6.11 we see the characteristics of a wind turbine: torque T versus speed Ω with the parameter wind velocity v; the relevant P_{max} line also indicated. A similarity is obvious; of course the rotating system will have a larger time constant than the PV system has.

Similar to MPP tracking in photovoltaic systems [Qua07], a searching algorithm can be established [Schi02] which checks variations of power P and rotational speed Ω at equally spaced time intervals, and adjusts the set-point speed by increments according to a suitable algorithm:

6.3 System Control

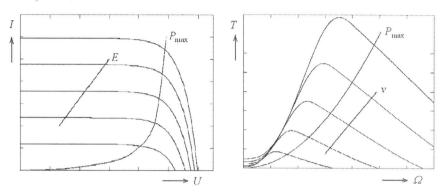

Fig. 6.11 Analogy of PV- and wind energy systems

Increase speed set-point if
$\Delta P > 0$ and $\Delta \Omega \geq 0$
or $\Delta P < 0$ and $\Delta \Omega < 0$

Decrease speed set point if
$\Delta P < 0$ and $\Delta \Omega \geq 0$
or $\Delta P > 0$ and $\Delta \Omega < 0$

The Ω command signal is used in the following active power control loop of a field-oriented control. Note that measuring the wind speed is not necessary.

Other publications describe variants of the method under the name of Hill-Climb Searching (HCS), e.g. [Wan04] who use an on-line training process and intelligent memory.

6.3.1.3 Multiple Input Control

Larger systems which normally feed into public grids receive their command values from the operation management. Controlled variables are normally active power P and speed n; in systems containing inverters also reactive power Q. Internal controlled variables can be, depending on system features, the pitch angle β and stator and rotor voltages and currents.

The control concepts are usually closed-loop schemes, containing one or more loops in cascade arrangement. Controllers are mostly of the PI-type. Functional diagrams are frequently used to describe the structure and calculate input and output quantities.

6.3.2 Control of Systems Feeding into the Grid

6.3.2.1 Functional Diagrams

Regarding systems for grid supply, the following block diagrams illustrate the basic control concepts.

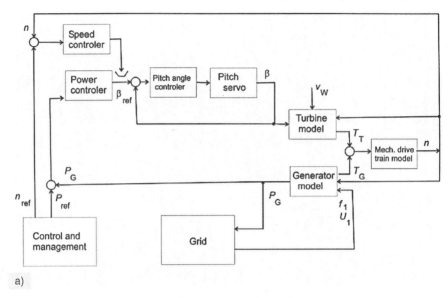

Fig. 6.12 Control scheme of a system for constant speed operation

Figure 6.12 is related to a constant speed WES with an induction generator directly feeding into the grid. The rotational speed is prescribed by the grid frequency with only small deviations. The system control provides reference values P_{ref} of power and n_{ref} of rotational speed. The power control is cascaded with the control of the blade angle β and its setting velocity. The speed controller is used to control the run-up procedure and to limit the power for wind velocities above rated value by acting on the pitch mechanism. Input values of the turbine model are wind speed, rotational speed and pitch angle; output is the turbine torque. The generators model inputs are grid frequency and voltage, together with speed n; output is the generator torque. Torques and rotational speed define the mechanical power values. Turbine and generator torques add up to the acceleration torque which is zero during steady state. The accelerating torque acts on the mechanical drive train model which in simplified form is an integrator, with the speed as output.

Figure 6.13 extends the block diagram to a variable speed system, the example assuming a fully-fed concept. While speed controller 1 limits the turbine power, speed controller 2 acts on the converter, supplying the set values for operation with e.g. optimum power. A reactive power (or power factor) control is usual but not shown in the figure. Variations of the figure are due when the generator is a doubly-fed induction machine, or when a synchronous generator requires additionally control of the excitation winding current.

A monograph on wind turbine control with special view on a proposed variant of gain scheduling techniques is in [Bian07].

6.3 System Control

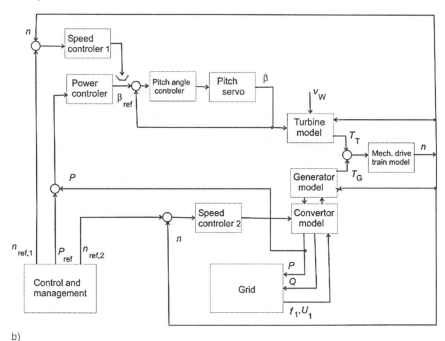

b)

Fig. 6.13 Control scheme of a system for variable speed operation

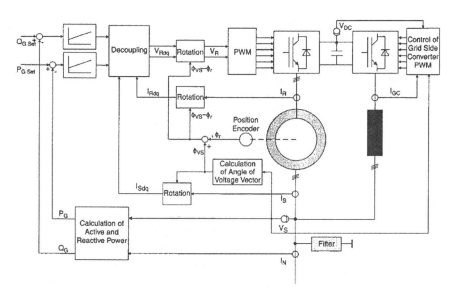

Fig. 6.14 Control scheme of a WES with doubly-fed induction generator

6.3.2.2 Vector Control

Vector control is an advanced method to control electrical machines [Quan08]. An example of vector control of a WES with doubly-fed asynchronous generator is illustrated in Fig. 6.14 [Mue02]. Generator power and reactive power are controlled independently by means of two PI-controllers. From measured values of terminal voltage and current actual values of the control variables P_G and Q_G are calculated and compared with set values $P_{G,set}$ and $Q_{G,set}$. Stator and rotor current components in dq-frame, i_{Rdq} and i_{Rdq}, are calculated from phase currents by means of vector

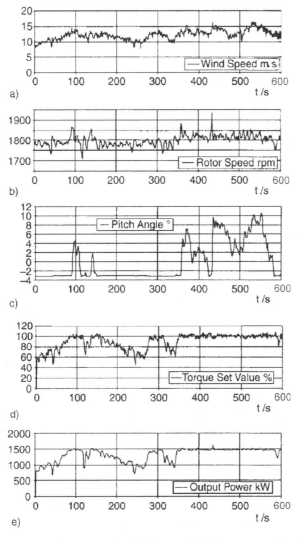

Fig. 6.15 Waveforms of the WES quantities during an example wind regime

rotation stages, with input values of the rotor position angle φ_r derived from a position encoder and of the calculated voltage vector angle φ_{VS}. In a decoupling stage the actuating variables for the rotor are calculated in dq-frame, V_{Rdq}, and subjected to a vector rotation to V_R in rotor frame. The converter is a VSI type with dc intermediate circuit. It is seen that V_R supplies the control signals to the rotor side inverter. The grid side inverter receives its PWM signals from a respective control.

The authors of [Mue02] have provided an example oscillogram of waveforms during a time interval of 10 min for a 1.5 MW wind system, nominal speed at 1800 min^{-1}, see Fig. 6.15. Note that below rated power the system is controlled from the electric side, preferably to optimum speed tip ratio; the power is then limited to rated value by pitch control.

Other control methods, known from industrial applications may also be applied. As an example [Sim97] describes a variable speed wind generation system where fuzzy logic principles are used for efficiency optimization and performance enhancement; a cage induction generator feeds power via a double-sided pulse width modulated converter into a utility grid or an autonomous system.

6.4 Basics of Operation Management

6.4.1 General

The operation management has to take care of a safe system performance. This requires continuous measuring of the quantities relevant for the system state. When specified limits are surpassed this must be checked. Faults must be detected and emergency routines initiated if necessary.

This includes monitoring of the states of operation at standstill, running up, waiting, performance under load and shutting down. Error states such as excess speed, impermissible temperatures and short-circuits must be detected and starting of countermeasures initiated. States of operation may be temporary or steady state.

6.4.2 States of Operation

System check
After activation of operation management the state of the system components is checked. Any errors must be cleared before the next states are commanded. System check is also made in operation before switching into other states.

Standstill
In the standstill state the rotor brake is on, rotors blades are in feathering direction (turned into the wind) while yawing is active, and the generator system is switched off.

Start up
When starting is commanded the turbine runs up on no-load, driven by the wind, until the speed has reached a suitable waiting state.

Waiting
In waiting state all components are in stand-bye. The electrical system is not connected to the POC. Conditions of switch-off on error and shut down are checked.

Run-up
In case of sufficient wind velocity the speed is brought to a suitable value, where after checking of generator and converter state, the electrical system is connected to the grid.

Load run
In part-load operation the system supplies power dependent on speed, usually under control on the electrical side. When reaching full load the pitch control becomes active. Overload operation within a limited interval is possible, as long as the thermal state of the system components permits.

Retardation
Retardation to a waiting state must be possible from any load or run-up operation. Blades can be brought in feathering direction, and electrical system can be disconnected.

Shut-down
Bringing the system to standstill must be possible from any operational state.

Disconnect on fault
Disconnection under fault condition is performed as a shut-down, but may be using steeper set-value ramps.

Disconnect on emergency
An emergency shut-down can be tripped either by the operation management or by the safety-system. Braking systems are used to decelerate as fast as possible. Any new start-up requires clearance by operator.

6.4.3 Grid Fault Reaction

In case of short-circuit a disconnection under fault should be made to prevent detrimental effects of the large currents on the electrical system. On the other hand the loss of load gives rise to turbine acceleration which must be limited by action of the pitch control. Similarly, when a short break happens in the grid, the normal action is to disconnect the system.

6.4 Basics of Operation Management

Short-circuit and fault ride-through properties required by transmission system operators as discussed in 7.5.1, however, can be met only if the WES is capable of continuing service without disconnection under voltage drop, as long as the voltage is above an established curve versus time, e.g. as shown in Fig. 7.10. Also the wind turbine system is expected to contribute a limited portion to the short current in the point of connection. These requirements set up by utilities have been a technical challenge to the wind system industry.

Chapter 7
Grid Integration and Power Quality

7.1 Basics of Grid Connection

7.1.1 General

The power curve of a wind turbine is the basic characteristic to describe its capability, see 2.4.3. Besides, the standards require the following operational actions to be tested:

Paralleling at cut-in wind speed,
Paralleling at rated wind speed,
Switching between generator stages, if applicable,
Switching of power factor compensation devices,
Service tripping from rated power,
Emergency tripping from rated power,
Unsymmetrical grid faults in each of the phases.

In order to obtain permission to parallel a wind energy system or a wind park to the grid, the transmission systems operators (TSO) require the declaration of technical properties. A main requirement is the limitation of voltage deviation caused by the wind systems at the point of common connection (PCC), for which 2% of nominal voltage is a commonly established limit. The short-circuit power at PCC is the crucial value for the permissible installed power ratings. Conventional regulations establish limits of frequency and voltage for normal operation. In the last years further-going guidelines were published which require wind parks to behave similar to conventional power stations regarding power factor, active power supply at frequency deviation and short-circuit capability in case of grid faults. Eventually harmonic and flicker limitation are power quality requirements.

7.1.2 Permissible Power Ratings for Grid Connection

WES generators are rated for low voltage, in Europe for 690 V or, less frequent, for 400 V. A machine transformer transforms the power to medium voltage level,

M. Stiebler, *Wind Energy Systems for Electric Power Generation*. Green Energy
and Technology. © Springer-Verlag Berlin Heidelberg 2008

e.g. 20 kV or regional standard voltage. The common connection point with the network operator may be on high-voltage level, which implies a grid transformer transforming from medium to high voltage level.

The apparent grid short-circuit power at the connection point is an important parameter in a project to erect a WES or a wind park, limiting the permissible power rating of the generators to be paralleled. In regions of rural character the distribution system, designed for consumer needs, may require power line reinforcement to allow economically reasonable WES installations at a chosen connecting point.

The voltage deviation occurring at the connecting point due to WES operation is determined by the short-circuit apparent power, together with the generator properties and the load conditions. The utilities set an upper limit for permissible voltage deviations in view of disturbing other grid-connected consumers.

The ratio of the short-circuit power and rated apparent WES or wind park power is called the connection factor; it is a criterion for acceptable rating of the generators to be installed in a wind park.

Usual grid regulation codes require that the voltage variation due to WES must not exceed 2% of nominal voltage at the connection point. Consequently, the maximum apparent power rating of the generator (or generators) S_{eV} in dependence of the apparent short-circuit power S_{kV} is required to comply with:

$$S_{eV} = \frac{P_{nG}}{\lambda} p_{10\,\mathrm{min}} \le \frac{S_{kV}}{k \cdot 50} \tag{7.1}$$

Reduction factor k in (7.1) takes account of rush-currents expected when connecting the system:

$k = 1$ for synchronous generators using a soft synchronizing method, and for converter coupled generators;

$k = 4$ for asynchronous generators, when switched on at 95% up to 105% of synchronous speed, without current limiting measures;

$k = I_a/I_N$ for asynchronous generators, when started as motors from the grid;

$k = 8$ in case I_a is unknown.

Asynchronous generators for direct coupling with the grid which are in use with ratings up to 2000 kW are switched on by thyristor soft starting devices (representing phase-controlled inverters) are considered $k \approx 2$. Systems above ratings of 1.500 kW are mostly coupled by means of converters, so that $k \approx 1$.

Generally a switching-current factor is specified to take account of voltage variations due to switching actions:

$$k_{\max} = I_{\max}/I_{nG} \tag{7.2}$$

A simple model to calculate voltage deviations at the connection point V is shown in Fig. 7.1a, where the grid is represented by an infinite bus and a short-circuit impedance. $\underline{U}_B$ is the infinite bus voltage, $\underline{Z}_k$ the impedance described by resistive and inductive component in series

$$\underline{Z}_k = R_k + j\,\omega\,L_k; \quad \psi_k = \mathrm{atan}(\omega\,L_k/R_k) \tag{7.3}$$

7.1 Basics of Grid Connection

Fig. 7.1 Model of short-circuit impedance between generator and grid and voltage variation

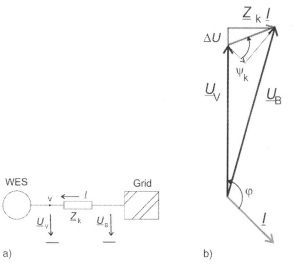

The WES is modeled by the generator symbol which includes line and transformer impedance components between terminals and connection point, referred to the relevant nominal voltage. In an example of a WES connected to medium voltage 20 kV, with a transformer 110/20 kV feeding into the distribution system, where $S_k = 120$ MVA referred to the 20 kV-side, the short-circuit impedance is given by:

$$Z_k = (2.1 + j\,2.6)\Omega\,;\quad |Z_k| = 3.342\ \Omega\,;\quad \psi_k = 51°$$

Current regulations require WES to declare a factor k_ψ, of influence for switching actions, which depends on the phase angle ψ_k of the short-circuit impedance $\underline{Z}_k$.

Consider a generator operating at an apparent power $\underline{S} = P + j\,Q$, the voltage variation at CP is approximately:

$$\Delta U_{aV} = R_k I \frac{P}{S} + X_k I \frac{Q}{S} = \frac{I}{S}(R \cdot P + X \cdot Q)$$

When the generator is operating at rated values, with current I_N and power factor $\cos\varphi_N$, the relative voltage increase becomes:

$$\Delta u_{aV} = \frac{\Delta U_{aV}}{U_N} = \frac{Z_k I_N}{U_N}\cos(\psi_{kV} - \varphi_N)$$

Note that the angles are in consumer system notation, where the sign of a lagging current (with respect to voltage) is positive, and the sign of the ohmic-inductive impedance $\underline{Z}_k$ is positive. In Fig. 7.1b an example vector diagram is shown which reflects the above equation under the condition that $|U_B/U_V - 1| \ll 1$. Note that in this case $U_B > U_V$ due to the case of a directly coupled asynchronous generator drawing its magnetizing current component from the grid. Theoretically, in the above example the voltage deviation would become zero when operating the generator at $\varphi = 141°$.

Using the generator and short-circuit apparent power values, the voltage deviation is calculated:

$$\Delta u_{aV} = \frac{\Delta U_{aV}}{U_N} = \frac{S_{nG} \cdot \cos(\psi_{kV} - \varphi)}{S_{kV}} \tag{7.4}$$

For specifications, the cos-function is set to 0,1, in case the calculated value is below.

7.1.3 Power Variation and Grid Reaction

Grid reactions created by wind energy systems appear in different forms. The wind-turbine accounts for:

- Power variations due to wind gusts
  Power variation is defined as the difference between the largest and the lowest power values during 8 periods within one minute. Single machine systems may encounter values between 0,6 and 0,9 rated power (see Fig. 7.2); in wind parks evening out occurs between systems, so that resulting variations are of magnitudes 0,25 up to 0,4 of rated power.
  Operation with as little variations as possible is intended by controlling the system on the turbine and/or on the generator side.
- Power variations due to tower shadow effects
  These periodic power variations can only be leveled out by a fast acting control.

Torque variations lead to dynamic torsional stress in the drive train which are especially relevant in systems for constant speed. Also unwanted are short-time power variations on the grid side.

The electrical system is subjected to:

- Switching operations
  Switching-on and off the generator may cause voltage fluctuations at the feed-in point. To limit inrush-currents in systems with directly coupled asynchronous machines phase-controlled thyristor-circuits are used, as mentioned. On the other

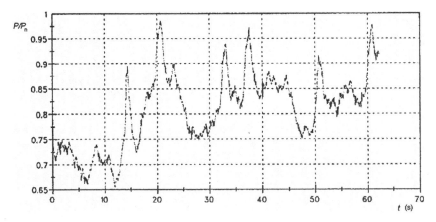

Fig. 7.2 Example of measured power variations

7.1 Basics of Grid Connection

hand the generator power should be controlled to zero before disconnecting (except emergency breaks).

- Reactive power
  Reactive power issues were discussed in chap. 3 for generators and in chap. 4 for phase-controlled inverters. The active factor $\cos \varphi$ reflecting the ratio of (fundamental) reactive and active power can be improved by compensation measures, such as fixed capacitor banks or controllable compensation devices.
- Flicker
  Voltage fluctuations of low frequency caused by power variations are called flicker, They give rise to lightness fluctuations of incandescent lamps and also of fluorescent lamps. In a band of around 1000 variations per minute they are experienced extremely inconvenient for the human eye.
- Harmonics due to inverters
  Grid-controlled inverters create current harmonics which give rise to voltage harmonics. The lowest order is determined by the pulse number of the inverter circuit. In six-pulse circuits, as in the three-phase bridge connection, these are the 5 and the 7 harmonic. In self-controlled PWM inverters the pulse frequency and its side-bands are prominent in the harmonic spectrum. Note that transistor inverters may be operated with pulse frequencies up to and exceeding 20 kHz, so that the audible components are above human hearing.

Voltage fluctuations in the connection point are influencing the grid voltage quality. Figure 7.3 shows voltage waveforms under typical distortions, as they occur due to WES or to other causes.

Standards and regulations set up by utilities describe the issues of power quality, and establish limits to be kept by WES feeding into the grid.

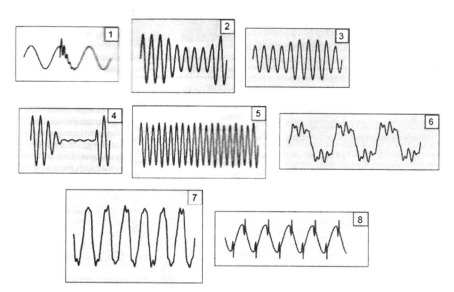

Fig. 7.3 Periodic and non-periodic voltage distortions. (1) oscillatory transient; (2) voltage sag; (3) voltage swell; (4) momentary interruption; (5) voltage flicker; (6) harmonic distortion; (7) voltage with interharmonics; (8) voltage with notches

7.2 Standard Requirements

7.2.1 Safety-Relevant Set Values

Systems with asynchronous or synchronous generators must be equipped with protective devices, with set values to allow adjustment of lower and upper limits of voltage and frequency. Recommended tripping values are given as follows (50 Hz rated frequency assumed):

	Limit	Tripping value
Voltage decrease	$0,70\,U_N$	$0,8\,U_N$
Voltage increase	$1,15\,U_N$	$1,06\,U_N$
Frequency decrease	48 Hz	49,5 Hz
Frequency increase	52 Hz	50,5 Hz

7.2.2 Reactive Power Compensation

The active factor cos φ, in literature often called the power factor, should be adjusted according to agreement. Usual recommendations require values between 0,9 cap. bis 0,8 ind. Use of capacitor banks is the conventional way to compensate inductive load. In grids where audio-frequency transmission devices are installed, the WES frequency response curve must be adapted by an appropriate inductive choke (see 7.6.3).

To obtain a specified cos φ value at the point of connection (POC), it may be necessary to provide a significantly higher reactive power in the wind park. Consider the example in Fig. 7.4 of a wind park with rated power 50 MW. A cos φ of 0,9 (capacitive) at the POC on the 110 kV HV side means that a 36,2 Mvar reactive power must be supplied on the wind turbine generator level, due to the inductive (magnetizing) currents for the LV/MV and MV/HV transformers, respectively. Capacitive contributions from assumed cable connections are only of minor influence.

7.2.3 Lightning Protection

It is necessary to provide wind turbines with lightning protection equipment. To this end non-metal blade tips are carrying interception apparatus, from where lightning currents are conducted to the hub by means of appropriate connectors. From the hub the lightning current is conducted to the metal tower construction and from there to the earthing system.

Since these measures are not sufficient to protect the electric and electronic systems, lightning protection zones (LPZ) are defined to create coordinated EMC-condition areas, see example in Fig. 7.5. The complete WES is situated in LPZ 0.

7.2 Standard Requirements

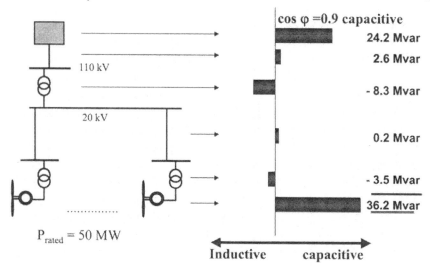

Fig. 7.4 Reactive power distribution example of a 50 MW wind park

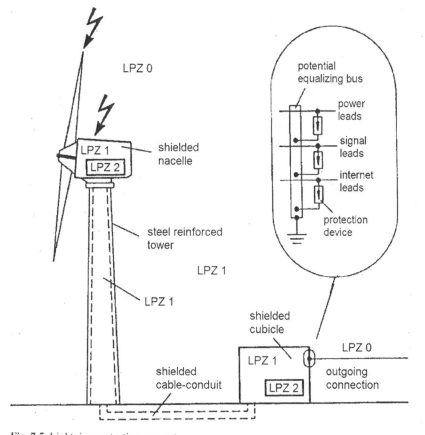

Fig. 7.5 Lightning protection concept

The shielded nacelle, the steel- or steel-reinforced tower and the cable-conduit form LZ 1. The electric control system is situated in the electro-magnetically shielded LPZ 2. Passing from one lightning protection zone to another, protection devices must be connected between the leads and the potential equalization bus.

7.3 System Operator Regulations

7.3.1 General

In view of the steady increase of WES installations, utilities have published additional regulations since 2001 [Eon03]. They require wind energy systems to contribute to a stable grid operation, and act similar to a conventional power plant. This means fault ride-through properties and the ability to supply reactive power in case of short-circuits.

First, the protective set-value requirements as reported in 7.4.1 are increased:

	Limit	Tripping time
Voltage decrease	$0,80\,U_N$	$3 \ldots 5\,s$
Voltage increase	$1,10\,U_N$	$<= 100\,ms$
Frequency decrease	$47.5\,Hz$	$<= 200\,ms$
Frequency increase	$51.5\,Hz$	$<= 200\,ms$

Other requirements are mentioned as follows:

- Starting from any operation point, the WES must be capable to reduce the power supply to a specified maximum value without disconnection from the grid; at least by 10% of rated power per minute.
- A wind park must be capable to operate at a power factor of between 0,975 ind. and 0,975 cap.
- In case of frequency deviations between 47, 5 and 51.5 Hz automatic breaking is not permissible; rather a specified power supply must be possible for a certain time interval.
- In case of grid faults with voltage sags no automatic breaking is permissible within a specified terminal voltage/time diagram.
- In case of short-circuit the wind parks must be capable of contributing to the short-circuit power for a certain time interval.

Active and reactive power are considered in the consumer (motor) coordinate system, where active and reactive power are assigned to the quadrants as in Fig. 7.6.

The following example guidelines are all for 50 Hz grids.

7.3 System Operator Regulations

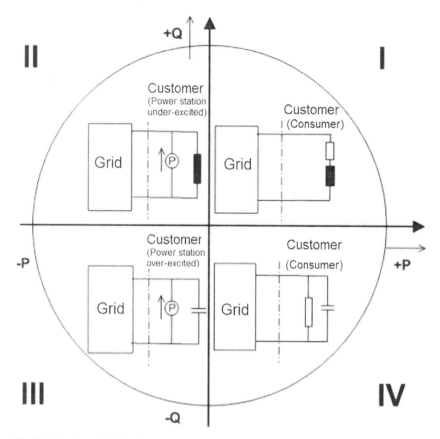

Fig. 7.6 Quadrant definition in consumer system

7.3.2 Active Power and Frequency

Utility guidelines require the WES to contribute to frequency control. Systems must be capable of supplying established minimum power, delivering their share to the power at underfrequency and overfrequency. In [Tra07] the limits are defined in a curve (see Fig. 7.7a), under which the system is not allowed to cut off. The Figure shows an example assuming a statics characteristic line of 5%. The power limits are applicable in a voltage interval of the high-voltage grid according to Fig. 7.7b where minimum time intervals before cutting is permissible are described.

Other grid codes define different variants of requirements regarding minimum sustained power delivery in dependence of frequency.

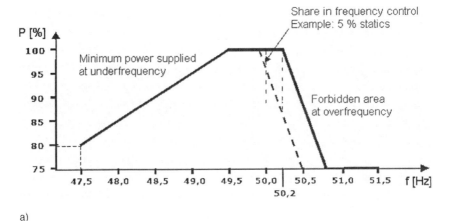
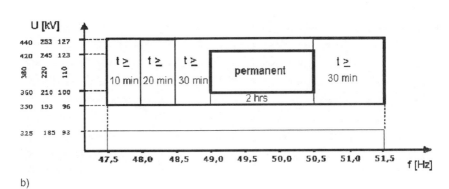

Fig. 7.7 Required active power capability of WES supplying a high-voltage grid

7.3.3 Reactive Power and Voltage

Utility guidelines also establish regulations regarding WES capability to supply reactive power and influence the power factor. Figure 7.8, taken from [VDN04], extends the requirement up to 31% of the installed apparent system power. The generator unit GU is required to serve in over-excited and under-excited condition; the Figure indicates the dependence on voltage in form of a statics line, allowing a hysteresis band. It is seen that the WES has to act as reactive power consumer at over-voltage, and has to deliver reactive power at under-voltage condition by acting as a capacitance.

Another description of the reactive power requirement in high-voltage grid, as taken from [VDN04, Tra07], is shown in Fig. 7.9. In the Figure, called variant 2, the voltage is connected with the power factor $\cos\varphi$ in a closed curve.

Variant 1 (not shown here) covers an area of $\cos\varphi$ between 0,975 inductive and 0,9 capacitive (instead of 0,95 and 0,925, respectively).

7.3 System Operator Regulations

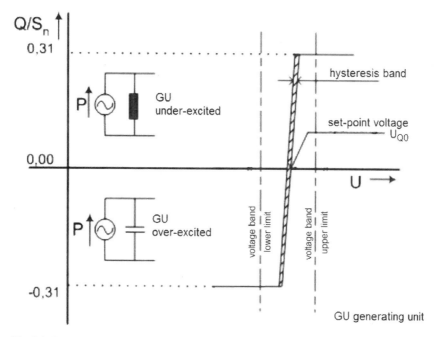

Fig. 7.8 Reactive power supply requirements

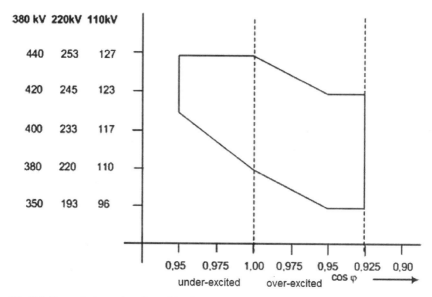

Fig. 7.9 Power factor assigned to grid voltage area

7.3.4 Short-Circuit and Fault Ride-Through

New grid codes require WES to contribute to supporting the voltage at the connection point, similar to a power station. In case of distance short-circuits, leading to a rest voltage of as low as 15%, the system is not allowed to trip within time intervals shown in Fig. 7.10 (50 Hz rated frequency, grid voltage level $\geq 60\,\text{kV}$), [VDN04]. This requirement covers cases where, due to successful stepping, the voltage is restored in times below 3 s. It is seen that service must be sustained in case the voltage has been restored to a permanent 80%. Utilities in other countries have established somewhat different curves, with variants of short-time rest voltage (down to zero, meaning terminals short circuited), slope of the curve and voltage to be sustained permanently.

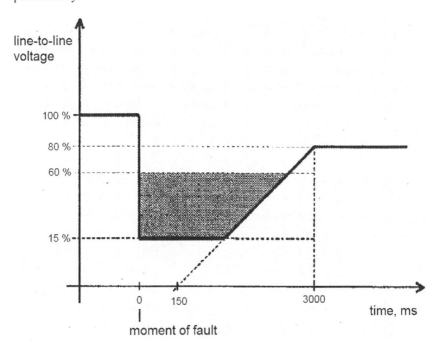

Fig. 7.10 Limiting voltage/time area excluding automatic tripping of wind parks

Also a WES should be capable of delivering a limited contribution to the short-circuit current. The current to be delivered at short-circuit is subject to agreement.

7.4 Power Quality

7.4.1 Harmonics

Harmonics are mainly generated in systems containing converters. Besides integer order numbers ν, non-integer order numbers μ appear in case of PWM controlled

7.4 Power Quality

converters. Permissible levels of harmonic currents are defined in IEC 61000-3-2. Utilities have established own permissible levels. Permissible values of harmonic currents are given in A/MVA, referred to short-circuit apparent power in the point of common connection. Regulations as defined by utilities in Germany for privately owned systems feeding into the grid, are given separately for low-voltage [VDW05], for medium voltage [VDEW98] and for high voltage [VDN04].

Table 7.1 gives limiting values for harmonic currents for generating systems feeding into LV, MV and HV grids, respectively.

Table 7.1 Permissible harmonic currents at connecting point

Order number ν, μ	Low voltage [VDW05] A/MVA	Medium voltage, 10 kV [VDEW98] A/MVA	High voltage, 220 kV [VDN04] A/GVA
3	4		
5/7	2,5/2	0,115/0,082	1,3/1,9
9	0,7		
11/13	1,3/1	0,052/0,038	1,2/0,8
17/19	0,55/0,45	0,022/0,018	0,46/0,35
23/25	0,3/0,25	0,012/0,010	0,23/0,16
$\nu > 25$	$0,25 \cdot 25/\nu$	$0,06/\nu$	$2,6/\nu$
$\nu = $ even	$1,5/\nu$	$0,06/\nu$	$2,6/\nu$
$\mu < 40$	$1,5/\mu$	$0,06/\nu$	$2,6/\nu$
$\mu > 40^*$	$4,5/\mu$	$0,18/\nu$	$8/\nu$

* integer & non-integer, in bandwidth of 200 Hz

An integral quantity characterizing the total harmonic distortion value is the THD. Regarding currents, harmonic components up to the order 40 are considered:

$$THD = \frac{\sqrt{\sum_{2}^{40} I_n^2}}{I_1} \qquad \text{where} \quad I_1 = \text{fundamental current} \qquad (7.5)$$

Electromagnetic compatibility (EMC) issues are addressed in an IEC standard [IEC60100], where IEC 61000-4-7 is a guide on harmonics and interharmonics measurement and instrumentation. For low voltage power systems IEC 61000-3-2 defines limits for harmonic current emissions.

7.4.2 Voltage Deviations and Flicker

Grid reactions generated by a WES must be limited in such a way that the service of equipment of other costumers and of utility devices is not disturbed. This is generally the case when the ratio of apparent powers $S_{kV}/S_{rA} > 500$, where S_{rA} is the rated power of the wind park.

Flicker is the subjective impression from light density variation of incandescent lamps. The maximum permissible values of voltage variations, their source, are a function of the frequency of their appearance. The conformity level $d = \Delta U/U_N$ is defined in dependence of their variations per minute. The values most inconvenient for the human eye are around 1000 changes per minute, corresponding to voltage variations of 8...10 Hz.

IEC 61000-3 describes the assessment of flicker values and their limitation. Defined are a short-time flicker strength P_{st} and a long-term flicker strength P_{lt}, based on a 10 min and 2 hour time-interval average, respectively. Figure 7.11, taken from IEC 61000-3-3, shows the curve for $P_{st} = 1$ for rectangular, equidistant voltage variations; it corresponds to the limit $d = \Delta u_{lim}$. Variants of the curve apply to other voltage variation waveforms.

Flicker disturbance factors A_{st} and A_{lt} are connected with the flicker strengths P_{st} and P_{lt} by the definitions

$$A_{lt} = (P_{lt})^3 \quad ; \quad A_{st} = (P_{st})^3.$$

In practice the P_{st} value can be determined by using empirical equations, using the flicker impression time t_f measuring deviations by means of

$$t_f = 2{,}3\, s \cdot (100 \cdot d \cdot F)^{3.2}, \text{ in s},$$

where

d is the maximum relative voltage change in % of rated voltage;

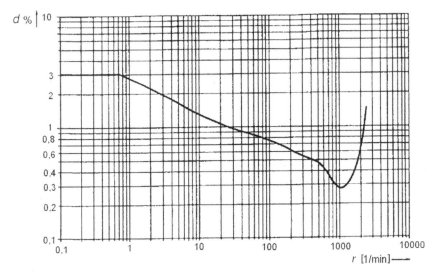

Fig. 7.11 Limiting curve of voltage variations per minute

7.4 Power Quality

$F \leq 1$ is the shape factor of the voltage change waveform.

The flicker strength P_{st} is the sum of the flicker impressions within a time interval T_p. When F is taken as 1, and T_p is chosen as 10 min (600 s), the short-time flicker strength is calculated

$$P_{st} = \left(\frac{\sum t_f}{T_p} \right)^{1/3,2} \tag{7.6}$$

According to EN 50160 the value $P_{st} = 1$ must not be exceeded in 95% of a week observation interval.

Flickermeters as described in IEC 61000-4-15 are used to record flicker, see Fig. 7.12. The block diagram indicates the input component 1, a quadratic demodulator 2, a band path and weighing filter 3, a variance estimator 4 and a statistic block 5.

Figure 7.13 shows some typical waveforms in the course of simulation of the lamp-eye-brain response.

The measuring devices contain a calculating algorithm, using weighing factors a_i to apply to cumulative frequency values $P_{i\%}$ of the power level spectrum to determine the short-time flicker strength:

$$P_{st} = \sqrt{\sum_1^5 a_i \cdot P_i} \tag{7.7}$$

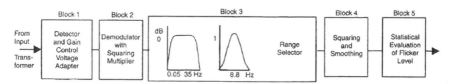

Fig. 7.12 Block diagram of a flickermeter

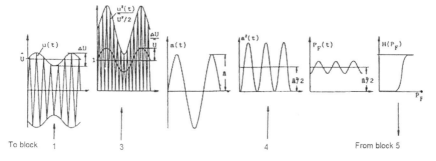

Fig. 7.13 Typical signal waveforms in a flickermeter

where

i	1	2	3	4	5
a_i	0,0314	0,0525	0,0657	0,028	0,080
P_i %	$P\,(0,1\%)$	$P\,(1\%)$	$P\,(3\%)$	$P\,(10\%)$	$P\,(50\%)$

The level values are recorded in a 10 min test series, as determined from measured quantiles of the relevant cumulative value, e.g. $P_{10\%}$. The long-time flicker strength follows with $N = 12$ for $12 \cdot 10$ min by cubic smoothing:

$$P_{\mathrm{lt}} = \sqrt[3]{\frac{1}{12} \sum_{1}^{12} P_{\mathrm{st.i}}^3} \tag{7.8}$$

An example of a cumulative power curve, taken from an arc-furnace (which is the main source of flicker) is shown in Fig. 7.11. The power levels P_i can be read from the curve $F(P)$, and the algorithm to determine P_{lt} applied.

Grid regulations require the long term flicker strength not to exceed $P_{\mathrm{lt}} = 0,65$ on the low voltage side. The code [VDEW98] for medium voltage level requires a tighter maximum of $P_{\mathrm{lt}} = 0,46$, corresponding to $A_{\mathrm{lt}} \approx 0,1$.

To assess the quality of one or more generating units relating to flicker-relevant voltage variations, a dimensionless flicker coefficient c is defined. It can be determined from tests under realistic operation conditions, alternatively declared by the manufacturer or a test institute. When c is known, the long term flicker strength is calculated by:

$$P_{\mathrm{lt}} = c\,\frac{S_{\mathrm{nG}}}{S_{\mathrm{kV}}}$$

Hence c is connected with the long-term flicker strength P_{lt}, and the ratio of the rated apparent power S_{nG} of a generating unit, and S_{kV} the short circuit power at the point of common connection. At further inspection, determination of relative voltage variations requires to take the relevant phase angle into account; in this case ψ_{kV} of the short-circuit impedance and φ_f, a flicker-relevant angle of the generator unit. Coefficient c is then determined from measured values of P_{lt} and φ_f of the generator:

$$c = P_{lt} \cdot \frac{S_{kV}}{S_{nG} \cdot |cos(\psi_{kV} + \varphi_f)|} \tag{7.9}$$

where

$$\varphi_f = \mathrm{atan}\left(\frac{\Delta Q}{\Delta P}\right)$$

Note that phase angles are counted from current to voltage, so that values are > 0 for inductive impedances. Fig. 7.14 illustrates the case of an asynchronous generator, connected to an infinite bus via a short-circuit impedance $\underline{Z}_k$; the circuit is similar to that in Fig. 7.2a. The current vector $\underline{I}_G$ for generator operation ends on the machine's current-locus circle of which a part is shown. The tangent in the working point is used for linearization; hence the phase angle of small current variations $\Delta \underline{I}$

7.4 Power Quality

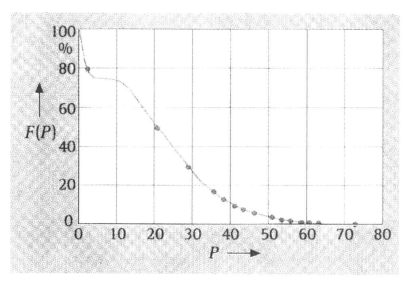

Fig. 7.14 Example from an arc furnace of a cumulative flicker power curve to determine P_{lt}

(length exaggerated in the Figure) is φ_f. In the example, the angle sum is approximately $(\psi_k + \varphi_f) \approx 80°$. Hence the projection of $\underline{Z}_k \Delta \underline{I}$ on $\underline{U}_V$ indicating the voltage variation is rather small. Note that c decreases with the cosine of the angle sum and can theoretically become zero. Regulations, however, prescribe to limit and set the cos-value $= 0,1$ if $|\cos(\psi_{kV} + \varphi_f)| < 0,1$.

Flicker coefficients vary with ratings and are generally larger for turbines with stall control than with pitch control. For a system where c is known, the long term flicker strength is calculated from equation (7.9) rearranged:

$$P_{lt} = c \cdot \frac{S_n}{S_k} |\cos(\psi_{kV} + \varphi_f)|$$

Noting that for a given WES the flicker coefficient is a function of the short-circuit impedance phase angle, the manufacturers are requested to declare c for selected values of the short-circuit impedance phase angle:

$$c(\psi_k) = P_{lt}(\psi_k) \cdot \frac{S_k}{S_n} \qquad (7.10)$$

Further, a grid-dependent switching current factor is defined which describes the influence of the system current on voltage variations:

$$k_{i\psi}(\psi_k) = \frac{\Delta U}{U} \frac{S_k}{S_n} \qquad (7.11)$$

In Table 7.2 characteristic values for selected systems (the same as in Table 5.1) are collated which are relevant for flicker and switching properties; also part-load power factors and permissible peak power values are given.

Table 7.2 Characteristics of selected systems (see Table 5.1) as determined from tests

Manufacturer	AN Windenergie GmbH	NORDEX AG	GE Wind GmbH	ENERCON GmbH	
System type	AN Bonus 33-2	N-50	1.5s	E-66	
Rated power	60/300	200/800	1.500	1.800	kW
Power factor λ $P/P_{NG} = 0,25/0,5/0,75/1$		0,99/1,0/1,0/1,0	1,0/1,0/1,0/1,0	0,99/1,0/1,0/1,0	
Peak power P/P_{NG} mom./1 min/10 min/		1,35/1,15/1,12	1.03/1,02/1,01	1,0/0,99/0,99	
Flicker coefficient c $\psi_k = 0/30/60/90°$		-/38/-/-	8/8/8/8	3/3/3/3	
Switch current factor $k_{i\psi}$ $\psi_k = 0/30/60/90°$		2,1/1,9/2,0/2,2	1,2/1,1/0,6/0,2	1,0/0.9/0,6/0,2	
$k_{i,max} = I_{max}/I_n$		2,3	1,1	1,0	

7.4 Power Quality

7.4.3 Audio Frequency Transmission Compatibility

Capacitors used for power factor compensation contribute to decrease the parallel resonance frequency of the grid. Consequently, the performance of audio frequency transmission equipment in use by the utility may be disturbed. Such transmission systems operate a signal frequencies of 250 up to 1500 Hz, and transmitter levels of 1% up to 4% of U_N. Hence the utilities require application of series inductors or trap circuits with the capacitances.

An example is Vattenfall in Berlin who operate a system using a signal frequency of 750 Hz. The company recommends an inductive compensation of 7%, i.e. the series inductors should be designed for 7% component of reactive power referred to the capacitive reactive power.

Figure 7.15 shows a simplified model of inductor-compensated capacitor. Its impedance is:

$$Z = Z_0 \left(\frac{\omega}{\omega_0} - \frac{\omega_0}{\omega} \right)$$

The inductive factor is defined as the ratio of reactive powers at fundamental frequency, which is equal to the voltage drop relation:

$$p = \frac{U_L}{U_c} = \omega_1^2 \cdot L \cdot C = \left(\frac{\omega_1}{\omega_0} \right)^2 = \frac{Q_L}{Q_C} \tag{7.12}$$

where

$\omega_1 = 2\,\pi\,f_1$ grid rated frequency,
$\omega_s = 2\,\pi\,f_s$ signal frequency,
$\omega_0 = 1/\sqrt{L\,C}$ resonant circuit eigenfrequency,
$Z_0 = \sqrt{L/C}$ characteristic impedance.

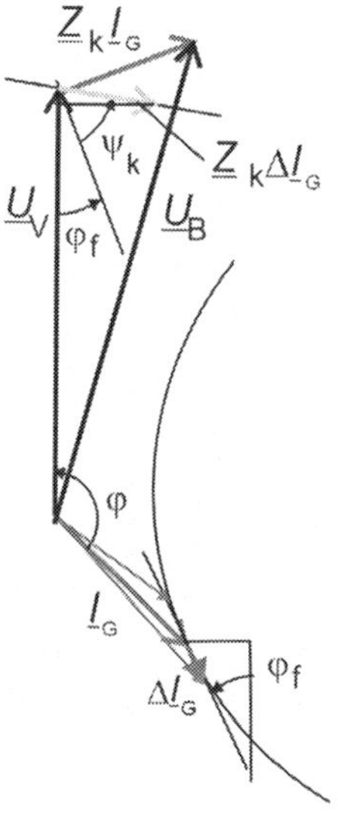

Fig. 7.15 Flicker-relevant voltage drop on a short-circuit impedance

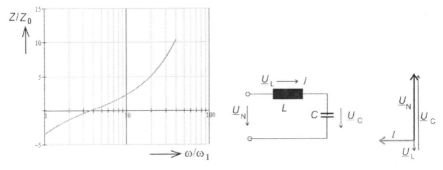

Fig. 7.16 Model of capacitive compensator with series inductor

In the example, for a required $p = 0.07$ and given $f_1 = 50\,\text{Hz}$, the inductor must be designed for $L \geq 70.92 \cdot 10^{-6}/C$ (in H when C is in F). The frequency response curves of the impedance referred to the capacitor impedance at ω_1, with and without series inductor are shown in Fig. 7.16; in the example the signal to grid frequency ratio is $\omega_s/\omega_1 = 15$. Note that in the model resistive components are neglected, and inductive/capacitive reactances are of positive/negative value.

7.5 Noise Emission

7.5.1 General

Sources of sound produced by mechanical systems such as wind turbines are:

– aerodynamic noise,
– magnetic noise,
– noise of bearings and slip-ring/brush contacts (if applicable).

These sources contribute in different magnitudes and frequencies to the sound emission.

The physical quantities most used in acoustics are expressed in logarithmic notation:

Sound pressure level $\quad L_p = 20 \cdot \lg(p/p_o)$, dB
$\quad$ with the reference $\quad p_o = 2 \cdot 10^{-5}$, Pa
Sound power level $\quad L_w = 10 \cdot \lg(P/P_o)$, dB
$\quad$ with the reference $\quad P_o = 1 \cdot 10^{-12}\,\text{W}$

The sound power from a source can be calculated from sound pressure levels, measured on a suitably defined surface, e.g. a half-sphere, put up above the machine under test. Test conditions and averaging procedure are specified in relevant standards. Using the measuring surface S, the relevant equation is:

$$L_w = L_p + 10\lg(S/S_o) \quad \text{where} \quad S_o = 1\text{m}^2$$

7.5 Noise Emission

The impression of sound on the human ear is a complex function of frequency and level. Generally, the ear is most sensitive for sound pressure near 1000 Hz, while the sensitivity decreases both to lower and higher frequencies. Experience from tests with numerous persons has led to the well-known Fletcher-Munson curves of equal loudness. On this basis weighting curves have been defined, called A to D. For practical use the A-weighting has become mandatory, in spite of the fact that this curve was originally meant for lower sound pressure levels of 40 dB. Levels weighted according to the A curve are given in dB(A).

For closer inspection, and to find single tones, sound measurements are made in octave or third-octave frequency bands.

7.5.2 Sound Emission by WES

During operation the audible noise produced by a WES is mainly determined by the aerodynamic sound of the blades, the principal components of which are due to the trailing edge noise and the noise from turbulent inflow. Noise components of other origin contribute only in a minor way. The generated noise depends on the wind velocity in hub height, the emission generally increasing with increasing load.

Pitch-controlled systems are characterized by power-dependent noise levels below rated power condition, and no further increase above rated wind velocity. On the other hand, in stall-controlled systems the noise emission increases further beyond rated power.

From experience it is known that the sound power level increases roughly 1–2,5 dB per increase of wind velocity by 1 m/s, under load up to rated condition, see Fig. 7.17.

For WES acoustic noise measurement techniques are standardized in IEC 61400-11 [IEC 61400]. Tests are made at specified wind velocities between 6 and 12 m/s,

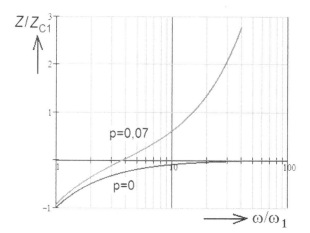

Fig. 7.17 Relative impedance of series resonant circuit

and up to the value corresponding to 95% of rated power. From the recorded data are determined:

- the A-weighted sound power level,
- the sound composition in form of third-spectrum,
- a tonal increment K_T in case of single tones,
- an impulse increment K_I in case of impulse containing sound.

Statistical values as a function of rated generator power are shown in Fig. 7.18. Typically, the sound power level produced by a wind generator system is determined $L_{WA} \approx 103\,\text{dB(A)}$, taking tonal and impulse increments into account.

The sound pressure level in the vicinity of wind parks depends, besides the wind velocity, on the systems types, the number of turbines and the location characteristics.

Examples of measured sound power levels of WES of different rating are shown in Fig. 7.19. An average of 103 dB(A) is typical for many systems at rated operation. Curves of equal sound pressure level are illustrated around a single system, and a wind park of 7 WES in Fig.7.20, respectively. The hub heights were 80 m, the measurements taken at a height of 5 m.

Noise has become a crucial issue in regional and local procedures of planning and licensing new wind parks. Established limiting values are considered by

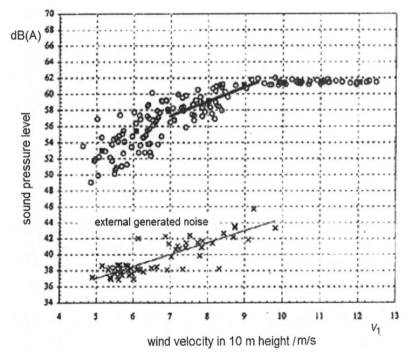

Fig. 7.18 Measured sound vs. wind velocity

7.5 Noise Emission

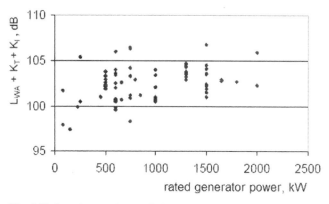

Fig. 7.19 Sound power levels of wind energy systems

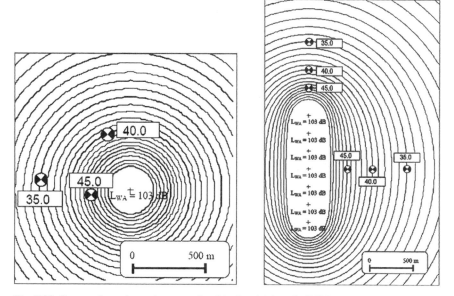

Fig. 7.20 Curves of equal sound pressure level in the vicinity of a WES

Table 7.3 Noise limits established by TA Laerm

Area declaration	limit (day)	limit (night)
Industrial area	70 dB(A)	70 dB(A)
Commercial area	65 dB(A)	50 dB(A)
Mixed and rural area	60 dB(A)	45 dB(A)
General housing estates	55 dB(A)	40 dB(A)
purely residential area	50 dB(A)	35 dB(A)
Hospitals, health cure and nursing areas	45 dB(A)	35 dB(A)

administrative and legal bodies. As an example reference is made of the German Technical Noise Regulation (Technische Anleitung Laerm). Dependent on the registered type of area maximum sound pressure values are specified. Especially the limits for night time are notable, see Table 7.3.

Practice shows that locations for the erection of wind parks must keep a minimum distance from residential areas of 600–800 m.

Chapter 8
Future of Wind Energy

8.1 Off-Shore Wind Energy Generation

8.1.1 General

Off-shore wind energy generation is considered as the techniques of dynamic development in the future. Existing off-shore wind parks in Europe such as Middelgrunden and Horns Rev have collected first practical experience. Studies suggest that worldwide offshore wind potential is larger than the electricity consumption [Ack02]. Currently a large number of wind parks are being in the planning and approval procedure.

Taking Germany as example, there are (as per end of 2007) a number of wind park projects licenced by the authorities in the exclusive economical zones of the North Sea (16) and in the Baltic Sea (6). The distances to coast are 13...100 km, water depths at park location are 15...40 m. Planning foresees an installed offshore power of 2900 MW until 2011, both in the North Sea and the Baltic Sea. Really large offshore wind parks are expected for the time after 2010.

8.1.2 Foundation

The foundation of the wind systems is the foremost technical issue, and will require a large portion of cost compared with on-land systems. Depending on the distance from shore of the location for erecting a WES and the water depth, several foundation methods have been investigated. [Mus05]

– Gravity foundation
  Gravity foundation features a flat base. It is stiff but heavy and has a larger footprint compared with the monopole. The construction must take care of stress due to ice, e.g. by providing an ice-breaking cone in height of the water surface level. The maximum water depth for which the solution is suitable is under discussion.

M. Stiebler, *Wind Energy Systems for Electric Power Generation*. Green Energy and Technology. © Springer-Verlag Berlin Heidelberg 2008

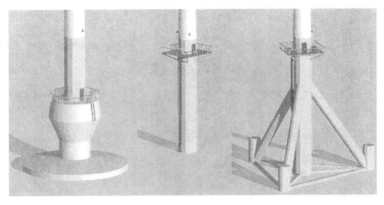

      Gravity base   Monopile   Tripod

Fig. 8.1 Foundation structures

- Monopile foundation
  This is the most common type for water depths up to 25 m. It has minimal footprint, but on the other hand features low stiffness. Important for the avoidance of resonance-induced dynamic oscillations is the knowledge of the structural frequency by design which may be a difficult task when there are uncertainties about the ground condition.

- Tripod
  For deeper waters, tripod support structures are being considered for offshore and coastal regions. Use of tripod foundation has been made down to 450 m water depth in the oil & gas industry, but with wind turbine systems there is little experience. Tripod constructions of which there are variants use a larger footprint.

- Floating
  Floating offshore wind turbines, e.g. on basis of a buoy concept, are a candidate mainly in shallow water. Currently they have the drawback of high cost, but future solutions may increase their potential.

  Figure 8.1 [OWE07] gives sketches of the foundation structure types.

8.1.3 Connection

Selection of the connection techniques between offshore wind parks and onshore stations is an important issue. Available solutions are HVAC connection and HVDC connection. Given the power rating, the choice of system has to take into account the wind park distance from shore defining the sea-cable length, the water depth and ground properties to assess investment cost, transmission loss and cable temperature rise, besides further concerns discussed in 8.1.3.

8.1.3.1 AC Connection

Considering a high-voltage AC connection it is mainly the voltage level of transmission which has to be decided. The problem with AC transmission is due to the capacitive cable impedance; consequently the current contains a reactive current component which gives rise to losses. A limiting cable length may be defined where, under idle running conditions, the magnitude of the (mainly capacitive) current amounts to the permissible cable current, without any active power being transmitted. At the same time a considerable voltage increase is observed at the receiving end.

Using the cable model illustrated in Fig. 8.2, the graph shows in a calculated example the normalized current at no-load I_{nl} (substantially a capacitive current) and the permissible active current I_{act}, referred to rated current I_r, versus normalized cable length. The limiting cable length l_{lim} is normally between 90 and 120 km. In practice inductive compensation devices are employed.

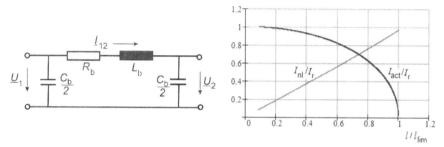

Fig. 8.2 Cable Π-model and limiting length

8.1.3.2 DC Connection

High-voltage DC connections are known in the classical form, where the stations on both ends of the cable are phase-controlled inverters using thyristors. Experience with this concept has been collected since first instalment 50 years ago, connecting Gotland with the mainland of Sweden. The modern version of HVDC transmission is using PWM-controlled inverters with IGBT semiconductors. Among other advantages this solution requires less space for the stations and allows power factor adjustment. Relating to short-circuit power the offshore and onshore AC sides are decoupled.

When comparing DC to AC transmissions, the AC terminal cost is less, while the line cost per cable length is less for the DC connection. Consequently AC connections will be of advantage for shorter distances, while the installation of DC lines is preferable above a certain distance. Figure 8.3 illustrates principle curves of investment cost over distance, where a rough estimate of the point of equal cost will be between 100 and 130 km. This is also where the efficiency of the DC transmission becomes better than for AC.

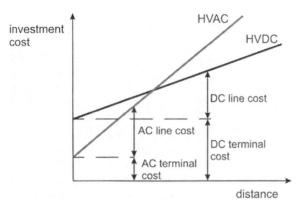

Fig. 8.3 Principal comparison of HVAC and HVDC connection cost

Model connection designs have been investigated. Examples based on a wind farm rating of 100 MW for both AC and DC connection are shown in Fig. 8.4 [Rei05]. The offshore grid is at 33 kV AC, star configuration; the undersea cable is operated at 150 kV AC or 150 kV DC, respectively, and the mainland AC side is also assumed at 150 kV [Bres07]. The cable connection may be made in one or two parallel undersea cables. Nominal voltage of 300 kV for undersea cables is under consideration.

In current technology the cable is using copper conductor, VPE insulation, a lead sheath and steel wire armour. Trends for future technology are superconducting and gas-isolated cables.

8.1.3.3 Comparison

When comparing AC and DC connection the differences affecting investment cost and losses should be noted, see Table 8.1.

Table 8.2 lists design features of modern HVDC connections as compared with the well-known classical concept. Consequently the advanced HVDC using active front-end voltage-source converters is considered the preferable solution.

8.1.4 Specific Issues and Concerns

In planning offshore wind farms several issues regarding environment, acceptance and safety have to be considered [Mus05].

– Nature conservation
  Regulations exist in view of the conservation of habitat for wild-living plants and animals. They establish natural reserve areas and define protected species of flora and fauna. In the European Union the topic is addressed by the FFH (Fauna,

8.1 Off-Shore Wind Energy Generation

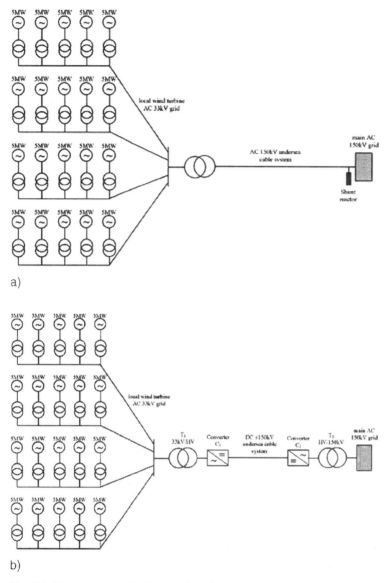

Fig. 8.4 Concepts of electrical connection of a wind farm 100 MW. (a) AC connection; (b) DC connection

176   8 Future of Wind Energy

Table 8.1 Basic comparsion of AC and DC connections

	AC	DC
Offshore installation	Lower space required	Larger space required
Conversion stations	Conventional; less cost	Converters required; higher cost
Cable	Larger loss at long connections; cable length limitation, higher cable cost	Lower cable cost; lower loss at long connections

Table 8.2 HVDC design features as compared with the classical concept

	Classical DC	Advanced-DC
Power semiconductors	Thyristors	GTOs, IGBTs, IGTCs
Switching frequency	Grid frequency (50, 60 Hz), phase control	Some kHz, PWM control
Filter devices	High	Low
Reactive power	Reactive power supply from grid required	Reactive power can be supplied to grid (up to converter rating)
Stability at no-load	No, at intermittent current	Yes
Maximum system ratings	Up to 1200 MVA, ± 500 kV	Up to 500 MVA, ± 150 kV

Flora, Habitat)-Guideline of 1992. A special guideline on wild-living birds was published already in 1979.

A point of discussion is in how far wind turbines are a danger for flying bird swarms. Available experience indicates that birds' collisions with turbines are scarce; they even suggest that rotating rotors are better seen and avoided than less than at standstill. It seems that HV overhead lines are more dangerous for birds then are wind turbines.

Taking the German North Sea and the Baltic Sea as example areas, an overview of the state of ecological research is found in [Koel06].

– Acceptance issues
  Wind turbines constitute an intervention in the landscape (viewshed) and give rise to noise emission and vibration, not principally different from systems on land. Other source of concern are of socio-economic nature, e.g. by fisheries.
– Physical and sea traffic issues
  Implications of electromagnetic fields due to undersea cable connection should be assessed. On the other hand, risks of collision for boating and ships' traffic require consideration. In a broader sense, hydrographic and coastal effects of WES installation on sea are a cause for inspection.

8.2 Power Integration and Outlook

8.2.1 Wind Energy in Power Generation Mix

In 2007 about 20 GW of wind power were installed worldwide. The first 5 of the contributing countries were USA (5240 MW), Spain (3520), China (3000), India (1700), Germany (1670)'; followed by France, Italy, Portugal, UK and Canada. Speaking of Germany, the portion of renewable energies in electricity consumption in 2007 was 14,8%, of which 7,2% came from wind systems.

The percentage installed power rating as well as of electric energy consumption is expected to increase in the next years. In this situation a view on future generation mix is appropriate.

8.2.2 Integration in Supranational Grids

Already now the high concentration of wind power in northern Germany produces power flows within the country and the neighbouring countries, and affects system stability and trading capacities. In medium-term additional reactive power compensators and reinforcement of networks will be necessary. To maintain system stability, fault ride-through capabilities must prevent the shedding of larger portions of wind power generation in case of grid faults.

For risk mitigation the European Wind Integration Study [EWIS07], with a view on the UCPTE network, proposes investigations for a time horizon 2015 on four levels:

- Political analysis
  Political support and priority for grid reinforcement and expansion
- Legal analysis
  Harmonization of rules; improvement of Grid Code requirements for wind power plants
- Market and business analysis
  Integration of balancing market; harmonization of regional markets; coordination by TSOs.
- Technical analysis
  Sharing of wind power forecast information; usage of large scale energy storage, systems of emergency control and demand side management; offshore integration

8.2.3 Outlook on 2020

The notion of Climate Change which is meanwhile generally accepted has intensified the efforts to develop and increase the share of renewables in energy consumption. It is obvious that wind energy plays a prominent role in all scenarios.

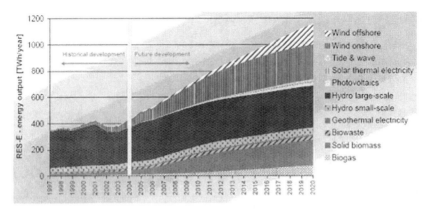

Fig. 8.5 EU electricity projection by 2020

In Europe the European Commission put forward an integrated energy/climate change proposal January 2007, addressing the issues of energy supply, climate change and industrial development. Renewable energy plays its role in the three sectors of electricity, heating and cooling, and transport [CEC08]. A plan was formulated calling for:

- 20% increase in energy efficiency
- 20% reduction in greenhouse gas (GHG) emissions
- 20% share of renewables in overall EU energy consumption by 2020
- 10% biofuel component in vehicle fuel by 2020

To achieve a 20% share of renewables by 2020 will require major efforts across all sectors of the economy and by all Member States. In the proposed EU Directive it is up to the Member States to decide on the mix of contributions from these sectors to reach their national targets.

In a Renewable Energy Technology Roadmap a projection was made of the composition of renewables for Europe up to 2020 [CEC07]. Figure 8.5 indicates that hydro, traditionally the main renewable energy form for electricity generation, will see almost no further increase. Wind energy will reach and surpass hydro, with offshore wind showing the largest increase rates. The so-called RES-E projection, based on 2004 values, estimates a factor of about 2.5 in consumption until 2020.

The European Wind Energy Association (EWEA) estimates that EU installed wind power capacity will reach 80 GW in 2010 and 180 GW by 2020. This would cover more than 12% of Europe's electricity demand. Offshore wind parks would reach installed power ratings up to 1000 MW, the size of conventional power plants. Turnover and employment in wind industry will reach values of importance for national economies; estimated is a turnover of $185 \cdot 10^9$ EUR between 2006 and 2020, and around 200000 direct and indirect employees in the EU-25 by 2020. [EREC07].

Annex A – List of Symbols

1. Turbine side

A	swept area, m^2
c_p	power coefficient
c_T	torque coefficient
D	rotor diameter, m
F	force, N
n	rotational speed, s^{-1}
v	wind velocity, m/s
λ	tip speed ratio
α	blade angle
β	pitch angle
ρ	specific air mass, kg/m^3

2. Generator and grid side

C	capacitance
$\cos \varphi$	active factor
f	frequency, Hz
I	line current, A
L	inductance
n	rotational speed, s^{-1}
P	power, loss, W
P_0	input power at no-load, W
P_1	input power, excluding excitation, W
P_2	output power, W
P_e	excitation circuit losses, W
P_f	excitation (field) winding losses, W
P_{fe}	iron losses, W
P_{fw}	friction and windage losses, W
P_C	constant losses, W
P_L	load losses, W

Q	reactive power, VA
R	winding resistance, Ω
Q	reactive power
s	slip, in per unit value of synchronous speed
S	apparent power
T	machine torque, $N \cdot m$
U	voltage, V
X	reactance, Ω
Y	admittance, Ω
$\underline{Z} = R + j \cdot X$	notation of a complex quantity (impedance as example)
$Z = |\underline{Z}| = \sqrt{R^2 + X^2}$	absolute value of a complex quantity (impedance as example)
Z	impedance, Ω
z_p	number of pole pairs
ϑ	load angle (synchronous machines)
λ	power factor
φ	phase angle
Φ	magnetic flux, Wb
ψ	short-circuit impedance angle
Ψ	flux linkage, Wb
σ	leakage factor
ω	pulsation, s^{-1}
Ω	angular speed, s^{-1}
η	efficiency
θ	temperature, °C
τ	time constant, s

Subscripts

Machine component

1, 2	primary, secondary
a	armature
e	excitation
f	field winding
ph	per-phase value
r	rotor
s	stator
syn	synchronous
w	winding
U, V, W	phase designations

Machine category

E	exciter
G	generator
M	motor

Annex A – List of Symbols

Operating condition

0	no-load
1	input
2	output
av	average, mean
B	base value
d	dissipated, direct current
el	electrical
i	internal
k	short-circuit
mech	mechanical
n	nominal
N	rated

Note: Units are SI-units

Annex B – List of Abbreviations

AC	Alternating current
CSI	Current source inverter
DC	Direct current
DSO	Distribution system operator
EDLC	Electric double-layer capacitor
EU	European Union
EWIS	European wind integration study
HV	High voltage
HVDC	High voltage DC
IEC	International Electrotechnical Commission
IEEE	Institute of Electrical and Electronic Engineers
LPZ	Lightning protection zone
LV	Low voltage
MPP	Maximum power point
PCC	Point of common connection
PMSM	Permanent magnet excited synchronous machine
POC	Point of connection
PV	Photo voltaic
PWM	Pulse width modulation
SEIG	Self-excited induction generator
TFM	Transversal flux machine
THD	Total harmonic distortion
TSO	Transmission system operator
UCTE	Union for the coordination of Transmission of Electricity
VSI	Voltage source inverter
WES	Wind energy system

Bibliography

[Ack02] Ackermann, Th., Leutz, R., Hobohm, J.: World-wide offshore potential and European projects IEEE Power Engineering Society, summer meeting 2002

[Ack05] Ackermann, Th. (ed.): Wind Power in Power Systems. Wiley, 2005, ISBN 0-470-85508-8

[Ale06] Alepuz, S. et al.: Interfacing renewable energy sources to the utility grid using a three-level inverter. IEEE Trans on Industrial Electronics, vol. 53, 2006, pp. 1504–1511

[Alep06] Alepuz, S. et al.: Interfacing renewable energy sources to the utility grid using a three-level inverter. IEEE Tryans. on IE, vol. 53, 2006, pp. 1504v–1511v

[Ars99] Arsory. A.B. et al.: Transient modeling and simulation of a SMES coil and the power electronic interface. IEEE Trans. on Applied superconductivity, vol. 9, 1999, pp. 4715–4724

[AWEA] American Wind Energy Association Fact sheets 2007 – A Record Year www.awea.org

[Bet26] Betz, A.: Windenergie und ihre Ausnutzung durch Windmühlen (in German). Vandenhoek und Rupprecht, Göttingen, 1926

[Bian07] Bianchi, F., de Battista, H., Mantz, R. Wind turbine control systems. Springer, 2007, ISBN 1-84628-492-9

[BRD04] Bundesrepublik Deutschland: Gesetz zur Neuregelung des Rechts der erneuerbaren Energien im Strombereich (EEG) (in German). 2004

[Bres07] Bresesti, P., Kling. W. et al.: HVDC connection of offshore wind farms to the transmission system. IEEE Trans. on EC, vol. 22, 2007, pp. 37–43

[Bul01] Bull, St.R.: Renewable energy today and tomorrow. Proc. IEEE, vol. 89, 2001, pp. 1216–1226

[BWE07] BWE Wind energy market 2007/2008. Bundesverband Windenergie e.V., 2007

[Cari99] Caricchi, F. et al.: Modular axial-flux permanent-magnet motor for ship propulsion drives. IEEE Trans on EC, vol. 14, 1999, pp. 673–679

[Carr06] Carrasco, J.M. et al.: Power electronic systems for the grid integration of renewable energy sources: A survey. IEEE Trans. IE, vol. 53, no. 4, 2006, pp. 1002–1016

[Cas92] Casacca, M.A., Salameh, Z.M.: Determination of lead-acid battery capacity via mathematical modelling techniques. IEEE Trans. EC, vol. 7, no. 3, 1992, pp. 442–446

[CEC07] Commission of the European Communities Renewable energy roadmap COM(2006) 848 final, Brussels, 2007

[CEC08] EU Press release Memo on the renewable energy and climate change package MEMO/08/33, Brussels, 2008-02-20

[Chap04] Chapman, St.: Electric Machinery Fundamentals, 4th Ed., McGraw-Hill, 2004, ISBN 0-07-246523-9

[Chat06] Chatterjee, J.K. et al.: Analysis of operation of a self-excited induction generator with generalized impedance controller. IEEE Trans. EC, vol. 22, no. 2, 2006, pp. 307–315

[Crot01] Crotingo, F. et al.: Huntorf CAES – More than 20 years of successful operation. SMRI Conf. paper, Spring 2001. Orlando/FLA

186   Bibliography

[DEWI] Wind turbine assessment DEWI, Wilhelmshaven, Services

[eia] Energy Information Administration (eia): World net generation of electricity by type, www.eia.doe.gov

[End08] Ender, C. Wind energy use in Germany – Status 31.12.2007 – (in German and English) DEWI, Wilhelmshaven, Febr. 2008

[enwi08] Grid energy storage (article) en.wikipedia.org/wiki/Grid_energy_storage

[Eon03] E.on Netz: Ergänzende Netzanschlussregeln für Windenergieanlagen (in German) E.on Netz 2003

[EREC07] European Renewable Energy Council Renewable energy technology roadmap up to 2020 Renewable Energy House, Brussels, 2007

[ewea] The European Wind Energy Association (EWEA): Wind power installed in Europe by end of 2006, www.ewea.org

[EWIS07] European wind integration study towards a successful integration of wind power into European electricity grids Final report. 2007

[Far06] Farret, A., Simoes, M.G.: Integration of alternative sources of energy, Wiley-IEEE Press, 2006, ISBN 0-471-71232-9

[Gas02] Gasch, R., Twele, J.: Wind Power Plants. James & James, 2002, ISBN 1-902916-38-7

[Gas07] Gasch, R., Twele, J.: Windkraftanlagen (in German). 5th ed., Teubner, 2007, ISBN 978-3-8351-0136-4

[GWEC07] GWEC Global Wind 2006 Report Global Wind Energy Council, 2007

[Hal96] Halpin, S. M. et al.: Application of double-layer capacitor technology to static condensers for distribution system voltage control. IEEE Trans on Power Systems, vol. 11, 1996, pp. 1899–1904

[Hau06] Hau, E.: Wind Turbines – Fundamentals, Technologies, Application, Economics, 2nd ed., Springer, 2006, ISBN 3-540-24240-6

[Hei05a] Heier, S.: Grid Integration of Wind Energy Conversion Systems, 2nd ed., Wiley, 2005, ISBN: 0-470-86899-6

[Hei05b] Heier, S.: Windkraftanlagen (in German), 4th ed., Teubner, 2005, ISBN 3-519-36171-X

[Hen97] Henneberger, G., Bork, M: Development of a new transverse flux motor, IEE Coll. on Topologies for Permanent Magnet Machines. 1997 (Digest 1997/090)

[Heu96] Heumann, K.: Grundlagen der Leistungselektronik (in German); 6th ed., Teubner, 1996, ISBN 3-519-46105-6

[Hin00] Hingoram, N.H., Gyugyi, L.: Understanding FACTs IEEE Press, New York, 2000, ISBN 0-7803-3455-8

[IEA06] International Energy Agency IEA Wind 2006 – Annual Report www.ieawind.org

[IEC60034] IEC IEC 60034 Rotating electrical machines, Part 1 (2004-04): Rating and performance, Part 2-1 (2007–09): Standard methods for determining losses and efficiency from tests, Part 4 (1985–01): Methods for determining synchronous machine quantities from tests. International Electrotechnical Commission, Geneva

[IEC60146] IEC IEC 60146 Semiconductor converters, Part 1: General requirements and line commutated converters (3 subparts), Part 2: Self-commutated semiconductor converters including direct d.c. converters ... and further part. International Electrotechnical Commission, Geneva

[IEC61000] IEC IEC 61000 Electromagnetic compatibility (EMC), Part 3: Limits – 3-2 (2005–11) Limits for harmonic current emissions (equipment input currents $<=$ 16 A per phase), 3-3 (2005–10) Limits – Limitation of voltage changes, voltage fluctuations and flicker in public low-voltage supply systems, for equipment with rated current $<=$ 16 A per phase, Part 4: Testing and measurement techniques – 4–7 (2002–08) General guide on harmonics and interharmonics measurements and instrumentation, for power supply systems and equipment connected thereto. 4–15 (2003–02) Flickermeter – Functional and design specifications International. Electrotechnical Commission, Geneva

[IEC61400] IEC IEC 61400 Wind Turbine Generator Systems, Part 1 (2007–03): Design Requirements, Part 2 (2006–03): Design requirements for Small Wind Turbines, Part 11 (2006–11): Acoustic noise measurement techniques, Part 12-1 (2005–12): Power performance measurements of electricity producing wind turbines, Part 21 (200-12). Measurement and assessment

Bibliography

of power quality characteristics of grid connected wind turbines International Electrotechnical Commission, Geneva

[IEC62428] IEC IEC 62428 Electric power engineering – Modal components in three-phase AC systems – Quantities and transformations International Electrotechnical Commission, Geneva

[IEEE112] IEEE Std 112-1996: IEEE Test procedures for polyphase induction motors and generators The Institute of Electrical and Electronics Engineers, New York

[IEEE115] IEEE Std 115-1995: IEEE Guide: Test procedures for synchronous machines, Part I: Acceptance and performance tests, Part II: Test procedures and parameter determination for dynamic analysis The Institute of Electrical and Electronics Engineers, New York

[IS106] Fraunhofer Institut System- und Innovationsforschung "Monitoring and evaluation of policy instruments to support renewable energy in EU member states" German Federal Environment Agency (UBA), 2006

[Kin04] Kinjo, T. et al.: Output levelling of wind power generation systems by EDLC energy storage. 30th Annual Conf. of IEEE Electronic Society, 2004, pp. 3088–3093

[Koe106] Köller, J., Köppel, J., Peters, W. (Eds.) Offshore wind energy – Research on Environmental Impacts Springer, 2006, ISBN 3-540-34676-7

[Kra02] Krause, P., Wasynczuk, O., Sudhoff, S.: Analysis of electric machinery, 2nd ed., Wiley-IEEE Press, 2002, ISBN 0-7803-1101-9

[Luo96] Luongo, C.: Superconducting storage systems: An overview. IEEE Trans. on Magnetics, vol. 32, 1996, pp. 2214–2223

[Moh95] Mohan, N., Undeland, T., Robbins, W.: Power Electronics, 2nd ed., Wiley, 1995, ISBN 0-471-58408-8

[Mue02] Müller, S., Deicke, M., de Doncker R.: Doubly fed induction generator systems. IEEE IAS Magazine, May/June 2002, pp. 26–33

[Mul00] Muljadi, E., Nix, G., Bialasiewicz, J.: Analysis of the dynamics of a wind-turbine water-pumping system IEEE

[Mus05] Musial, W.: Wind powering America – Potential for the United States National Renewable Energy Laboratory, 2005

[Muy07] Muyeen, S.M. et al.: Comparative study on transient stability analysis of wind turbine generator system using different drive train models. IET Renewable power generation, vol. 1, no. 2, 2007, pp. 131–141

[Nirg01] Nirgude, G., Tirumala, R., Mohan, N., A new, large-signal average model for single-switch DC-DC converters operating in both CCM and DCM. IEEE PESC 2001, vol. 3, 2001, pp. 1736–1741

[Ohy07] Ohyama, K., Arinaga, S., Yamashita, Y.: Modeling and simulation of variable speed wind generator system using boost converter of permanent magnet synchronous generator Conf. EPE'07

[Ok195] Okla, O.: Permanenterregter Ringgenerator für kleine Windkraftanlagen (in German), Verlag Dr. Köster, Berlin, 1995

[OWE07] Offshore wind energy OWE, Expert guides, Technology of OWE. www.offshore windenergy.org

[Qua05] Quaschning, V.: Understanding renewable energy systems. Earthscan, 2005, ISBN 1-844-07128-6

[Qua07] Quaschning, V.: Regenerative Energiesysteme (in German), 5th ed., Hanser, 2007, ISBN 3-446-40973-4

[Quan08] Quang, N.P., Dittrich, J.-A.: Vector control of three-phase AC machines Springer, 2008, ISBN 978-3-540-79028-0

[Ram07] Ramacumar, R. et al.: Introduction to the special issue on wind power (contains 27 papers on energy development & power generation). IEEE Trans. on EC, vol. 22, 2007, pp. 1–3.

[Rei05] Reidy, A., Watson, R.: Comparison of VSC based HVDC and HVAC interconnections to a large offshore wind farm. IEEE Power Engineering Soc. General Meeting 2005, vol. 1, 2005 pp. 1–8

[Sal92] Salameh, Z. M. et al.: A mathematical model for lead-acid batteries. IEEE Trans. EC, vol. 7, 1992, pp. 93–97

Bibliography

[San06] Saniter, C.: Frequency-domain modelling of voltage source converters and doubly-fed induction machines facing distorted ac power networks VDE Verlag Berlin, 2006, ISBN 978 3-8007-2989-0

[Schi02] Schiemenz, I., Stiebler, M.: An optimum searching algorithm for control of a variable speed wind energy system OPTIM, Brasov/Romania, 2002

[Schm56] Schmitz, G.: Theorie und Entwurf von Windrädern optimaler Leistung (in German) Z. d. Universität Rostock, 5 (1955/56)

[Schn02] Schneuwly, A. et al.: BOOTSCAP Double layer capacitors for peak automotive applications, AABC-02, 15pp

[Stie00] Stiebler, M., Dietrich, W.: Design criteria for large permanent synchronous machines, Proc. ICEM'2000 Helsinki, pp. 1261–1264

[Schu06] Schulz, D.: Integration von Windkraftanlagen in Energieversorgungsnetze – Stand der Technik und Perspektiven für die dezentrale Stromerzeugung Berlin, VDE Verlag, 2006, ISBN 3-8007-2949-0

[Sim97] Simoes, M.G., Bose, B.K., Spiegel, R.J.: Fuzzy logic based intelligent control of a variable speed cage machine wind generation system. IEEE Trans. on Power Electronics, vol. 12, 1997, pp. 87–95

[Slo03a] Slootweg, J.G., de Haan, S., Polinder, H., Kling, W.L.: General model for representing variable speed wind turbines in power system dynamics simulations. IEEE Trans. PS, vol. 18, no. 1, 2003, pp. 144–151

[Slo03b] Slootweg, J.G., Polinder, H., Kling, W.L.: Representing wind turbine electrical generating systems in fundamental frequency simulations. IEEE Trans. EC, vol. 18, no. 4, 2003, pp. 516–524

[Sou01] Sourkounis, C. et al.: Autonomous network-control and power conditioning in decentral power supply systems with high share of fluctuating energy sources. 2nd Nat. Renewable Energy Congress, Athens, March 2001

[Sun05] Sun, T., Chen, Z., Blaabjerg, F.: Flicker study on variable speed wind turbines with doubly fed induction generator. IEEE Trans. EC, vol. 20, no. 4, 2005, pp. 896–905

[Svech06] Svechkarenko, D. et al.: Analysis of a novel transverse flux generator in direct-driven wind turbines IR-EE-EME 2006, p. 424ff.

[Tra07] Verband der Netzbetreiber Transmission Code Netz- und Systemregeln der deutschen Übertragungsnetzbetreiber (in German) Verband der Netzbetreiber-VDN, Berlin, 2007

[VAC] VAC Vacuumschmelze Vacodym, Vacomax – Catalogue VAC, Hanau/Germany, 2002

[VDEW98] Vereinigung Deutschere Elektrizitätswerke Eigenerzeugungsanlagen am Mittelspannungsnetz (in German) VWEW Frankfurt M, 2nd Ed. 1998

[VDN04] Verband der Netzbetreiber EEG-Erzeugungsanlagen am Hoch- und Höchstspannungsnetz (in German) Verband der Netzbetreiber-VDN, Berlin, 2004

[VDW05] Verband der Elektrizitätswirtschaft Eigenerzeugungsanlagen am Niederspannungsnetz (in German) VWEW Frankfurt M, 2005

[Wan04] Wang, Q., Chang, L.: An intelligent maximum power extraction algorithm for inverter-based variable speed wind turbine systems IEEE Trans. PE, vol. 19, 2004, pp. 1242–1249

[WEC04] World Energy Council: 2004 survey of energy resources. Elsevier, J. Trinnaman, A. Clarke, Editors, 2004, ISBN-10: 0-08-044410-5

[Weh86] Weh, H., May, H.: Achievable force densities for permanent magnet excited machines in new configurations Proc. Int. Conf. on Electrical Machines (ICEM), Munich, 1086

[Wha05] Whaley, D., Soong, W., Ertugrul, N. Investigation of switched mode rectifiers for control of small-scale wind turbines. IEEE IAS Annual Meeting 2005, vol. 4, pp. 2849–2856

Index

A.c. power controllers, 70
A.c./a.c. inverters, 56
Accelerating (motoring) torque, 123
A.c./d.c. inverters, 56, 57
Active front-end inverter, 64–66
Active power and frequency, 155–156
Active-stall, 21
Admittances, 41
Aerofoil theory, 14
Air-gap torque, 122
Armature short-circuit time constant, 133
Armature winding, 43
Asynchronous machine models
  in $\alpha\beta$–coordinates, 120–123
  in field-oriented components, 124–125
  modal component model, 123–124
  small deviations from steady state, 126–127
  transient model, 125–126
Asynchronous machines (AM), 29, 40, 148
  models, *see* Asynchronous machine models
  power conversion in, 30
  T-model circuit, 31–32
Audio frequency transmission compatibility, 165–166
Axial field machines, 49–50
Axial flux generators, 96

Battery-loaders, 70, 84, 107
Betz theory, 15
Bipolar transistors, 61
Breakdown-torque, 33

Choppers, *see* D.c./d.c. inverters
Clarke transformation, 117, 119–120
Combined generation, systems in
  diesel generation, combination with, 103–104

renewable power sources, combination with other, 104–106
Compressed air storage devices, 77–78
Conformity level, 160
Connection factor, 148
Constant speed, 85
Converters
  control, 94
  with intermediate circuits, 67
  modeling, 133–135
CSI, *see* Current-source inverter (CSI)
Current-source inverter (CSI), 57, 60–61, 67
Cyclic converters, 67

Danish concept, 52, 82, 83
D.c./a.c. inverters, 56
D.c./d.c. inverters, 57, 67–70
Dimensionless flicker coefficient, 162
Direct driven generators, 48, 51, 95
Double-layer capacitors, 75
Doubly-fed asynchronous machine, 52, 93–94
Drag coefficient, 14–16

EDLC, *see* Electrochemical double-layer capacitors (EDLC)
Electrical angle, 44–45
Electrical energy storage, 75–76
Electrical equipment
  conventional, 55
  energy storage devices, 71
    electrical energy storage, 75–76
    electrochemical energy storage, 71–75
    mechanical energy storage, 76–79
  power electronic converters, 55–57
    a.c. power controllers, 70
    converters with intermediate circuits, 67
    d.c./d.c. choppers, 67–70

external-commutated inverters, 57–61
self-commutated inverters, 61–67
Electrical shaft, 84
Electrical supply, wind energy contribution to
governmental regulations, 8–9
installed power, 3–7
technical standardization and local issues,
7–8
Electric power transmission, 84
Electrochemical double-layer capacitors
(EDLC), 75
Electrochemical energy storage, 71–75
Electromagnetic torque, 125
Electronic synchronous machine, 106
EN 50160, 161
Energy density of pump storage systems, 77
Energy storage devices, 71
electrical energy storage, 75–76
electrochemical energy storage, 71–75
mechanical energy storage, 76–79
$E'p$, 129
European Wind Integration Study (EWIS), 4,
177
EWIS, see European Wind Integration Study
(EWIS)
External-commutated inverters, 57–61

Fast-running turbines, 13
Flicker, 151
Flickermeters, 161
Floating offshore wind turbines, 172
Flux-building component, 125
Flywheel storage devices, 78–79
Force, 15
Fortescue transformation, 118

Gamma-equivalent model, 42
Generators, 29
asynchronous machines
grid operation, 34–38
model assumptions, 31–33
operation at given stator current, 39
operation at given stator voltage, 33–34
performance equations and equivalent
circuits, 31–39
per unit representation, 38
principles of operation, 30
reactive power compensation, 40–41
self-excited operation, 41–43
comparison, 52–53
synchronous machines
complex locus and vector representation,
45–46
model assumptions, 44

operation at given passive load, 46–47
operation at given stator voltage, 44–45
performance equations and equivalent
circuits, 44–48
permanent magnet excitation, 48
principles of operation, 43
unconventional machine types, 48–52
Glide ratio, 14
Gravity foundation, 171
Grid connection, basics of, 147
Grid-dependent switching current factor, 163
Grid fault reaction, 144–145
Grid integration and power quality
grid connection, basics of, 147
noise emission, 166–167
sound emission by WES, 167–170
permissible power ratings for, 147–150
power quality
audio frequency transmission compatibility, 165–166
harmonics, 158–159
voltage deviations and flicker, 159–164
power variation and grid reaction,
150–151
standard requirements
lightning protection, 152–154
reactive power compensation, 152
safety-relevant set values, 152
system operator regulations, 154–155
active power and frequency, 155–156
reactive power and voltage, 156–157
short-circuit and fault ride-through, 158
Grid reaction, power variation and, 150–151
Grid regulations, 162
Grid-supplied a.c. drives, 40

Harmonics, 158–159
Harmonic currents, Permissible levels of, 159
High temperature superconductors (HTSC), 76
High-voltage AC connections, 173
High-voltage DC connections, 173–174
HTSC, see High temperature superconductors
(HTSC)
Hybrid model, 134

IEC, see International Electrotechnical
Commission (IEC)
IEC 61000-3, 160
IEEE, see The Institute of Electrical and
Electronics Engineers (IEEE)
IEEE standards, 29
IGBT, 61
Inductive components, 122
Inductive factor, 165

Index

Inductor, 43
pole flux, 129
Infra-noise, 7
Institute of Electrical and Electronics
Engineers (IEEE), 201
International Electrotechnical Commission
(IEC), 7
Inverter performance, 150–151

Kirchhoff's law, 36
Kramer system, super-synchronous, 90–93

Lead-acid batteries, 71–74
LIBERTY, 96
Lift/drag ratio, see Glide ratio
Lightning protection, 152–154
Lightning protection zones (LPZ), 152–154
Limiting cable length, 173
Linear capacitor voltage function, 42
Linear network theory, 87
Load
angle, 45
torque deviation, 126–127
Low temperature superconductors (LTSC), 76
LPZ, see Lightning protection zones (LPZ)
LTSC, see Low temperature superconductors
(LTSC)

Machines
with excitation winding, 96–97
performance, 30
with permanent magnet excitation, 97–102
Matrix converters, 67
Maximum power point (MPP) tracking, 138
Mechanical energy storage, 76–79
Modal component transformations, 115–116
γ-model, 87
Modern wind turbines, theory of, 5
Modular magnet machines, 52
Modulation ratio, 63
Monopile foundation, 172
MOSFET, 61
MPP, see Maximum power point (MPP)
MULTIBRID, 96
Multi – inertia system, 135

NaS-batteries, 74
Nature conservation, 174–176
NdFeB magnet material, 97
NiCd-batteries, 74
NiMH-batteries, 74
Noise emission, 166–167
sound emission by WES, 167–170

Operation management, basics of, 143
states of operation
disconnect on emergency, 144
disconnect on fault, 144
load run, 144
retardation, 144
run-up, 144
shut-down, 144
standstill, 143
start up, 144
system check, 143
waiting, 144
Ossanna diagramme, 35
Overmodulation, 64

Park' theory, 44
Park transformation, 117, 120
PCC, see Point of common connection (PCC)
Performance and operation management, 115
basics of, 143
grid fault reaction, 144–145
states of operation, 143–144
system component models
asynchronous machine models, 120–127
converter modeling, 133–135
modeling the drive train, 135–137
model representation, 115–120
synchronous machine models, 127–133
system control
control of systems feeding into the grid,
139–143
multiple input control, 139
optimum control by model characteristic,
138
optimum control by MPP tracking,
138–139
Permanent magnet, 129
excitation, 48, 97–102
machines, 97
Permanent magnet excited synchrous machines
(PMSM), 107
systems with, 107–111
Photovoltaic (PV) generator, 106
Pitching mechanism, 20
PMSM, see Permanent magnet excited
synchrous machines (PMSM)
Point of common connection (PCC), 147,
159, 162
Point of connection (POC), 145, 152
Pole axis, 44
Power behaviour of induction machine, 37
Power electronic converters
converters with intermediate circuits, 67
d.c./d.c. choppers, 67–70

external-commutated inverters, 57–61
self-commutated inverters, 61–67
Power factor, 152
Power-invariant and power-variant transformation, 116–117
Power limitation concepts, 22
see also Pitching mechanism; Stall mechanism
Power quality
audio frequency transmission compatibility, 165–166
harmonics, 158–159
voltage deviations and flicker, 159–164
Power ratings, 147–150
Power-variant transformation, 121
Power variation and grid reaction, 150–151
Pulse width modulation (PWM), 63, 67, 143
inverters, 151, 173
PWM, see Pulse width modulation (PWM)

Rayleigh distribution, 85
Reactive power
compensation, 152
current-source inverter, 60–61
and voltage, 156–157
voltage-source inverter, 66–67
Rectifiers, see A.c./d.c. inverters
Reference-component-invariant transformations, 116
Reference energy yield, 26
Reference flux linkage, 38
Reference inductance, 38
Renewable energies
sources of, 1–2
sources of electrical energy production, 2–3
wind and solar energy, 3
RES-E projection, 178
Resistance, 74

Safety-relevant set values, 152
Sankey-diagram, 37
Secondary batteries, 74–75
SEIG, see Self-excited induction generator (SEIG)
Self-commutated inverters, 61–67
Self-commutated voltage-source converter, 66
Self-excitation, 41
Self-excited induction generator (SEIG), 42–43, 111–112
Short-circuit, 144, 145
and fault ride-through, 158
Slip, 31–32
Slipringless excitation, 96

SMES, see Superconducting magnetic energy storage (SMES)
Soft-start devices, 70
Solar energy, 1
Space phasors, transformation into, 117–118
Spring coefficient of shaft segment, 135
Stall mechanism, 20–22
Stand-alone systems
induction generator, systems with, 111–113
PMSM, systems with, 107–111
small systems in the kW range, 106–107
Superconducting magnetic energy storage (SMES), 76
Superconducting storage devices, 76
Super-synchronous cascade, 60
Super-synchronous Kramer system, 90–93
Synchronous angular speed, 38
Synchronous generator inductors, 44
Synchronous machines (SM), 29, 43, 95
complex locus and vector representation, 45–46
in grid-operation, 45–46
model assumptions, 44
models, see Synchronous machine models
operation at given passive load, 46–47
operation at given stator voltage, 44–45
performance equations and equivalent circuits, 44–48
permanent magnet excitation, 48
principles of operation, 43
unconventional machine types, 48–52
Synchronous machine models
machine with damper cage, 131–132
machine with fieldwinding on the rotor, 127–129
modeling the drive train, 135–138
reactance operators and frequency response, 132–133
for small deviations from steady state, 130–131
transient model, 129–130
System component models
asynchronous machine models, 120–127
converter modeling, 133–135
modeling the drive train, 135–137
model representation, 115–120
synchronous machine models, 127–133
System control
control of systems feeding into the grid, 139–143
multiple input control, 139
optimum control by model characteristic, 138
optimum control by MPP tracking, 138–139

System operator regulations, 154–155
  active power and frequency, 155–156
  reactive power and voltage, 156–157
  short-circuit and fault ride-through, 158

Terminal voltage, 39
TFM, *see* Transversal flux machines (TFM)
ThD, *see* Total harmonic distortion
Three-phase full wave bridge inverter, 61–62
Thyristor bridge inverter, 57–60
Torque, 14–16, 19–20, 33, 39, 122–123, 125, 126–127, 138
  building component, 125
  variations, 150
Total harmonic distortion (THD), 159
$\alpha\beta0$ transformation, 121
Transformation, non-rotating frame, 118
Transformations and reference frames, 118–120
Transient inductor e.m.f., 129
Transmission system operators (TSO), 147, 177
Transversal flux machines (TFM), 51
Tripod, 172
TSO, *see* Transmission system operators (TSO)
Two-axis theory, *see* Park' theory

Unconventional designs, 48
  axial field machines, 49–50
  transversal flux machines, 51
Unconventional machine types
  direct driven generators, 48
  unconventional designs, 48
    axial field machines, 49–50
    modular magnet machines, 52
    transversal flux machines, 51
    variable reluctance machines, 52
Uninterruptible power supplies (UPS), 79
UPS, *see* Uninterruptible power supplies (UPS)

VACODYM 655 AP, 97–98
Variable reluctance machines, 52
VENSYS, 95
Voltage deviations and flicker, 159–164
Voltage source inverter (VSI), 33, 57, 61, 66, 67, 134
Voltage-source PWM inverter, 62–64
VSI, *see* Voltage source inverter (VSI)

Water power, 1
Water pump storage, 76–77
WES, *see* Wind energy systems (WES)
Wind and solar energy, 3
Wind energy, future of
  capability of, 85
  off-shore wind energy generation, 171
    connection, 172–174
    foundation, 171–172
    issues and concerns, 174–176
  power integration and outlook
    integration in supranational grids, 177
    outlook on 2020, 177–178
    wind energy in power generation mix, 177
Wind energy contribution to electrical supply
  governmental regulations, 8–9
  installed power, 3–7
  technical standardization and local issues, 7–8
Wind energy systems (WES), 81
  properties of, 100–102
  systems for feeding into grid, 84–85
    asynchronous generators in static cascades, 86–94
    commercial systems, 103
    induction generators for direct grid coupling, 85–86
    synchronous generators, 95–102
  systems for island operation
    stand-alone systems, 106–113
    systems in combined generation, 103–106
  systems overview, 81–82
    systems feeding into the grid, 82–83
    systems for island supply, 83–84
    wind pumping systems with electrical power transmission, 84
Windmills, 11
Wind turbines, 11
  power characteristics and energy yield
    annual energy yield, 26–27
    control and power limitation, 20–22
    system power characteristics, 23–25
    wind classes, 22
  torque, 138
  wind energy conversion
    forces and torque, 14–16
    power conversion and power coefficient, 11–14
  wind regime and utilization
    power and torque characteristics, 19–20
    power distribution and energy, 18–19
    wind velocity distribution, 17–18

CPSIA information can be obtained at www.ICGtesting.com
233786LV00011B/1/P